Environmental Compliance:
A Web-Enhanced Resource

T0314343

Environmental Compliance: A Web-Enhanced Resource

Edited by
Gary S. Moore, Dr. P.H.

CRC Press
Taylor & Francis Group
Boca Raton London New York

CRC Press is an imprint of the
Taylor & Francis Group, an **informa** business

A TAYLOR & FRANCIS BOOK

CRC Press
Taylor & Francis Group
6000 Broken Sound Parkway NW, Suite 300
Boca Raton, FL 33487-2742

First issued in paperback 2019

© 2001 by Taylor & Francis Group, LLC
CRC Press is an imprint of Taylor & Francis Group, an Informa business

No claim to original U.S. Government works

ISBN-13: 978-1-4200-4318-1 (hbk)
ISBN-13: 978-0-367-38718-1 (pbk)
Library of Congress Card Number 00-063790

Library of Congress Cataloging-in-Publication Data

Moore, Gary S.
    Environmental compliance : a web-enhanced resource / Gary S. Moore.
        p. cm.
    Includes bibliographical references and index.
    ISBN 1-56670-520-7 (alk. paper)
      1. Environmental law --United States. 2. Environmental sciences--Computer network resources. I. Title.

KF3775 .M66 2000
344.73′046—dc21   00-063790

Visit the Taylor & Francis Web site at
http://www.taylorandfrancis.com

and the CRC Press Web site at
http://www.crcpress.com

AUTHOR

Dr. Gary S. Moore

Dr. Moore is a faculty member in the Department of Environmental Health Sciences in the School of Public Health and Health Sciences at the University of Massachusetts. He has extensive publications (over 75) in the area of health effects from environmental insults and has been awarded nearly $1 million in grants and contracts while at the university.

He has written, illustrated, and developed several complete environmental courses with ancillary slides, transparencies, and workbooks. He recently (1997-1999) worked with several of his classes in EnvHl 565 and three graduate students in particular to write and illustrate a comprehensive textbook titled "Living with the Earth," which has received national attention, and has been selected as one of *Choice* Magazine's OUTSTANDING ACADEMIC BOOKS of 1999 judged on distinguished scholarship. *Choice* is a publication of the Association of College and Research Libraries, a division of the American Library Association, a not-for-profit organization. Some of the graduate students have indicated that it was among the most meaningful educational experiences of their graduate education. Dr. Moore also worked with the class to simultaneously create and produce a web site for the book and course to allow students access to ancillary course materials while also featuring many interactive features. This is considered to be a leading edge technology. The present text titled "Environmental Compliance - A Web-Enhanced Resource" is also a class-involved, web-enhanced book featuring class members from Dr. Moore's EnvHl 562 class. This is a bolder initiative than the previous text designed to give his entire class an opportunity to become published. This will also serve as the basis for an online course as part of the initiative to develop a degree-based online program for Public Health.

Most recently, Dr. Moore took a lead position in writing and submitting a grant proposal to eCollege on behalf of the School of Public Health, Isenberg School of Management, the School of Nursing, and the Division of Continuing Education. The grant was recently awarded at $207,000 and represents the highest award that eCollege has made to the hundreds of the competitive applications. He is now focusing efforts on bringing online, web-based degree programs to the university community in cooperation with the Division of Continuing Education. His course is the first online course in this effort, and will serve as model for courses to follow. The course has been rated by eCollege staff as the "best" online course in their experience. Dr. Moore is committed to bringing the very latest technologies to the classroom and beyond for an exciting and animated learning environment.

Dr. Moore was also instrumental in obtaining grants to develop the "Guidebook for Massachusetts Boards of Health," which was released in June, 1997. Presently, he is developing and providing certification and training courses for boards of health throughout the state of Massachusetts and is working with the Department of Environmental Protection, the Massachusetts Department of Public Health, and the Massachusetts Association of Health Boards.

He has received several awards for his efforts in this area during 1998 and 1999 including recognition from the Massachusetts Senate, the Massachusetts House of Representatives, The Massachusetts Association of Health Boards (MPHA), and the Robert C. Huestis/Eric W. Mood award from the New England Public Health Association (NEPHA). The certification and training programs for boards of health have been hugely successful. Since 1995 Dr. Moore has assisted MAHB in establishing the groundwork for this first-in-the-nation program. This project is a strong foundation for further collaboration between the university and MAHB, providing practical educational opportunities to people serving every community in the Commonwealth.

CO-AUTHORS

Jeff Bagg

Jeff has completed an associate's degree in the liberal arts from Springfield Technical Community College. He is entering his senior year at the University of Massachusetts, Amherst, where he plans to obtain a bachelors degree in natural resource conservation with a minor in environmental science. His concern for the environment and its natural resources stems from ecological disturbing local events and strong respect for the environment. His future plans lean toward a career in policy development and educating people about the importance of the environment.

James O. Bailey

James Bailey is a supervisor at Mohawk Plastics Company in Bernardston, Massachusetts, where he coordinates safety and environmental compliance regulations matters. He has an extensive work history with Dart Industries, Mobil Chemical Co., and BASF Corporation. James has an A.A.degree in liberal arts from Springfield Technical Community College (STCC). He is working on a B.S. in environmental safety management at the University of Massachusetts, Amherst.

Katherine Brossard

Katherine Brossard has recently graduated from the University of Massachusetts, Amherst with a B.S. degree in economics of public policy for natural resources. She spent her junior year studying in Edinburgh, Scotland. In her free time she enjoys music, travel, and wildlife.

Charles Cicconi, M.D.

Charles is a medical doctor in the U.S. Navy completing an advanced degree in public health at the University of Massachusetts, Amherst.

Andrea Depatie

Andrea Depatie is currently an environmental health major at the University of Massachusetts Amherst. She plans to continue on to graduate school to pursue a career in public health.

Kevin Doherty

Kevin Doherty will be graduating from the University of Massachusetts, Amherst with a B.S. degree in environmental science. He currently resides in Massachusetts where he enjoys surfing, hiking and fishing.

David R. Gillum

David graduated from the University of Nevada, Las Vegas, in 1997. He received his B.S. degree in chemistry. He has good knowledge of environmental health and safety programs with more than five years work experience. David received his M.S. in environmental health science from the University of Massachusetts, Amherst. David is pursuing a career in the field of environmental health and safety.

Evan Hutchinson

Evan graduated from Springfield College, where he received his B.S. degree in human services. He is currently pursuing a graduate degree in environmental health sciences at the University of Massachusetts, Amherst. His work experience includes working for a major oil company in the Caribbean before returning to school. He also has a background in automotive technology and enjoys the world of motor racing and helping friends trouble-shoot automotive problems.

Ian Cambridge

Ian is an environmental health major. He is currently enrolled as an undergraduate at the University of Massachusetts, Amherst. He hopes to pursue a career in medicine and continue on to Jefferson Medical School after his graduation.

Sean Jobin

Sean is an undergraduate at the University of Massachusetts, Amherst. He is expected to graduate in December of 2000 with a B.A. degree in resource economics. He has experience in managing a landscaping service and has contributed some time to the Massachusetts Public Interest Research Group. Sean currently lives in Uxbridge, Massachusetts. He enjoys fishing, hiking and outdoor recreational activities.

Sherri McGloin

Sherri earned her associate degree in applied science, and certification as an operating room practitioner from Southern Maine Technical College. She received her B.A. in public health from the University of Massachusetts, Amherst..

Marc A. Nascarella

Marc received his B.S. degree in environmental sciences with a concentration in engineering from the Military College of Vermont, Norwich University, Northfield, Vermont. He is pursuing an M.S. in environmental health sciences, with a concentration in environmental toxicology, from the University of Massachusetts School of Public Health and Health Sciences. His work experience includes employment at major engineering firms and medical centers as well as military service in the U.S. Air Force. Marc's current research is focused on effects of low-level chemical and radioactive stress on longevity.

Michael A. Pepe

Michael received his B.S.degree in human biology from American International College. While at American International College, Michael gave a series of seminars concerning environmental pollution. Michael is currently enrolled in the Masters of Public Health program at the University of Massachusetts, Amherst. Michael enjoys outdoor recreational activities and currently resides in Massachusetts.

Julie Vander Ploeg

Julie will receive her B.S. degree in environmental science and resource Eeconomics from the University of Massachusetts, Amherst. She is a three-year member of the nationally ranked women's crew team. She has worked as an intern at the Sunderland, Massachusetts Board of Health and as a CAP inspector for the University of Massachusetts Environmental Health and Safety Office.

Travis Veracka

Travis received his B.S.degree in environmental science from the University of Massachusetts, Amherst. His work experience includes RCRA compliance issues and site inspection He was a member of the varsity baseball team for the University of Massachusetts for four years. Currently, Travis lives in New England, where he enjoys hunting, fishing, and the outdoors.

Rachael D. Weiskind

Rachael received her B.A.degree in psychology from the University of Tennessee. She earned an M.A. degree in sport and exercise science and completed an interdisciplinary specialization in geriatrics from Ohio State University. Rachael then earned her MPH from the School of Public Health and Healt Sciences at the University of Massachusetts, Amherst. Her work experience includes peer consultation, fitness and health appraisal, wellness coordination and program development, health and safety compliance, and air and noise sampling methodologies. Rachael's interests and hobbies include spending time with her family (specifically the family pet), cooking, and running.

Maggie Coe Wood

Maggie is currently working on her B.A. degree in architecture and urban studies at Smith College. She has been actively involved in environmental issues, and is interested in environmentally conscious architectural design. She is looking forward to going to Denmark next year where she will continue her studies in architecture and environmental policy. Currently, Maggie lives in Massachusetts where she enjoys outdoor activities and playing the flute.

Michael Wood

Michael received his B.S. degree in environmental science with a concentration in environmental policy from the University of Massachusetts, Amherst. He has earned his certification under the OSHA 40 hour, 29 CFR 1910.120 Hazardous Waste Operations and Emergency Response (HAZWOPER) program. He has experience in writing health and safety plans for industrial situations. He also is experienced in writing ASTM Phase I Environmental Site Assessments.   Michael has obtained skills in groundwater sampling for petroleum and/or hazardous materials. He is interested in furthering his education and concentrating his studies on hydrogeology. Michael lives in upstate New York, and enjoys fishing, hiking, and the outdoors.

PREFACE

This book is intended to be a college level textbook for courses in environmental health and environmental sciences and engineering. It is intended to be a useful resource for practicing environmental professionals. It is also highly recommended for members of boards of health, health officers and health inspectors, and citizens who are members of proactive environmental groups such as Sierra Club, Public Interest Research Groups, Clean Water Action, Environmental Defense Fund, and many others.

This book contains 15 carefully narrated chapters on environmental regulations and compliance. The contents are derived from an environmental course Dr. Moore instructs at the University of Massachusetts, Amherst titled EnvHl 562, Environmental Regulations and Compliance. This book incorporates both the traditional concepts associated with environmental regulations and compliance and new issues of environmental justice and ISO 14000. Careful attention is paid to presenting a balanced view with ample access to reputable web sites that provide up-to-the-minute information in a rapidly expanding regulatory world.

There is a dedicated Web site associated with this book and the related university courses. You may access this web site at *<http://www-unix.oit.umass.edu/~envhl567>*. Within this Web site are a host of chapter specific links selected to help you maintain a current knowledge base.

ACKNOWLEDGMENTS

To my wife Lucille for her unwavering support during my long periods of isolation in the cellar office and in what my children refer to as "the bat cave." She accepted my dedication to the project and kept me fed and clothed in the interim. I especially give thanks to students in my Spring 2000 EnvHl 567 class who took on the enormous task of writing assigned chapters for this effort and they succeeded wonderfully.

CONTENTS

A SYSTEM OF LAWS

1

Sean Jobin

HOW A BILL BECOMES A LAW [1,2]

Any federal law starts as a bill and goes through a system of checks and balances. The bill is first introduced into the House and the Senate. These bills then pass by a subcommittee where they are reviewed and may or may not gain support. Almost 90% of bills fail to make it through this process. Bills that are recommended are brought through hearings where the committee makes comments and options. The committee meets to discuss the bill and to finally vote on it. If the committee in the Senate votes for the bill, it is sent to the full chamber where it is again debated and voted on. Before a bill is sent to the House, it must go to the rules committee where the bill receives a time limit for debate and they decide whether to allow floor amendments. Riders are sometimes attached to bills that do not have any relationship to the original bill. The riders tend to accompany popular bills so that they might be approved along with the primary bill.

The House and Senate usually pass a bill with many differences. The bills with differences must go into a conference so that the House and Senate may resolve these differences. Once the House and Senate agree on the same version of the bill, it is sent to the President. The president may sign or veto the bill. If the bill is vetoed then it will take a two-thirds majority vote in the House and Senate to override the President's veto.

A SYSTEM OF ENVIRONMENTAL LAWS [1]

There is a system of guidelines, executive orders, regulations, and statutes that in the past thirty years have become a system of environmental laws. All environmental laws originate from: the U.S. and state constitutions, federal, state, and local ordinances, regulations published by federal, state, and local agencies, presidential executive orders, court decisions that have interpreted these laws and regulations, and common law. Environmental law uses all of the laws in the legal system to reduce, prevent, and punish actions which have damaged or threatened the environment, public health, and its safety.

Almost all environmental laws use eight compliance obligations or regulatory approaches. These obligations are:

Discharge and waste control
Process controls and pollution prevention
Product controls
Regulation of activities
Notification requirements
Safe transportation requirements
Response and remediation requirements
Compensation requirements[1]

Notification Requirements [1]

When there is a release of hazardous wastes and pollutants all of the appropriate authorities and public must be advised and notified as soon as the spill exists. Many environmental impacts and dangers can result from violations and accidents due to waste spills.

Controls and Pollution Prevention [1]

There are at least three protective actions that can regulate hazardous wastes and they include:

• Minimizing waste generation
• Reducing the quantities produced and released
• Preventing the release of pollutants into the environment

Each state has different pollution prevention plans. To find examples of these programs you can go to *www.state.ma.us/dep/com*. Here you are able to search by region, research new and ongoing programs, and read about daily news releases.

Controls on Products [1]

Product design, usage, and packaging standards have been made to minimize solid waste generation and disposal problems associated with toxic and hazardous substances. The laws have been set to lower risk to human health and the environment. The standards provide information on solid and hazardous waste management practices, and waste water treatment activities. There is also a summary of recycling and equipment laws that are used when managing waste and wastewater treatment.

Regulating Activities [1]

We regulate activities to give protection for endangered species, ecosystems, and all natural resources. Some activities that are constantly regulated are: construction, mining, harvesting trees, and extracting natural resources like oil. All of these activities must be controlled to prevent irreversible harm to the environment. You can obtain a profile of every state agency's requirements at *www.govinst.com/pubscatalog/products/685.html.com*. This is also an internet guide that serves many purposes.

Safe Transportation Requirements [1]

Laws are intended to minimize the risks of hazardous wastes that are being transported by air, land, or trucks. These strict laws are made to reduce the risk of spills or accidents that might occur during the transportation of hazardous materials. The protection of the environment and public health is the main concern behind the transportation laws.

Response and Remediation Requirements [1]

There are laws that regulate the cleanup of pollutants and hazardous wastes that have been released into the environment. These laws also are intended to identify those who are responsible for the cleanup and who will pay the costs for that cleanup. In some states there is a full-service environmental contractor that specializes in emergency oil spill response, hazardous and toxic waste contamination. They can be contacted 24 hours a day by going to *www.ampol.net.com.*

Requirements for Compensation [1]

All responsible parties must pay for all cleanup costs and can be held liable for any damages that hurt the environment or any surrounding private property. Compensation can be awarded if public assets are damaged by spills.

FEDERAL STATUTES [3,4]

The Clean Air Act, Clean Water Act, and many other environmental acts are federal statutes that establish federal and state regulatory programs. The programs allow states to enact and enforce laws that meet federal minimum standards. The states must also meet the regulatory goals that have been established by Congress. Many states have taken over the regulatory programs in their areas. States are the main permitting and enforcing authorities and can be subject to federal help only when the enforcement from the states is not effective or is lacking. The regulations the states enforce are usually more stringent than the federal laws Each state may differ somewhat in the content and enforcement of these laws.

Executive Orders

Since 1970, many presidents have enacted many executive orders that require federal facilities to help protect the environment and to comply with statutes and policies that involve protection of the environment. A number of these Presidential Executive Orders include:

- Executive Order 11514: Protection and Enhancement of Environmental Quality
- Executive Order 11990: Protection of Wetlands
- Executive Order 12372: Intergovernmental Review of Federal Programs
- Executive Order 12844: Federal Use of Alternative Fueled Vehicles
- Executive Order 12856: Federal Compliance with Right-To-Know Laws and Pollution Prevention Requirements
- Executive Order 12873: Federal Acquisition, Recycling, and Waste Prevention

- Executive Order 12902: Energy Efficiency and Water Conservation at Federal Facilities

Executive Orders 12856 and 12873 were issued by President Clinton on August 3, 1993 and October 20, 1993. Both these orders have an enormous impact on the federal agencies.

Executive Order 12856: Federal Compliance With Right-To-Know Laws and Pollution Prevention Requirements

This order was meant to establish compliance of federal agencies with the Emergency Planning and Community Right-to-Know Act of 1986 and the Pollution Prevention Act of 1990. It was intended to insure all federal agencies minimize their waste to the maximum extent possible, the amount of toxic chemicals that enter any waste stream and the environment. All of this must be done through source reduction. Storage, treatment, or dumping of waste must be done in a way that protects the environment and public health.

The head of each federal agency must comply with the provisions set in EPCRA. These responsibilities are providing information needed for the Local Emergency Planning Committee (LEPC) to prepare and revise local Emergency Response Plans and to keep up-to-date Material Safety Data Sheets (MSDS).

All federal agencies must meet these requirements and file annual progress reports to the administrator starting on October 1, 1995. In this report the agency must describe the process it followed when trying to comply with all parts of the order and the pollution reduction requirements.

Executive Order 12873: Federal Acquisition, Recycling, and Waste Prevention

President Clinton also issued this order. The order requires the head of each executive agency to incorporate waste prevention and recycling in all daily work and operations. The agencies must use cost-effective procedures and programs when purchasing products and services.

Each agency is required to set a goal for solid waste prevention and a goal for recycling. These goals must be met by 1995. Once the agency has set the goal it must be submitted to the Federal Environmental Executive within 180 days of the date. Each federal agency must set annual goals to maximize the amount of recycled products they purchase. For more information on environmental laws visit *www.E.P.A.SummaryofEnvironmentalLaws.com.* At this site they provide you with explanations of many major environmental laws and what kind of an impact they have had in the U.S.

OTHER TYPES OF LAWS CONTRIBUTING TO THE SYSTEM OF ENVIRONMENTAL LAWS

State Water Protection Laws [3,5]

Almost all states have a detailed permit program for ground water protection. The federal government has not adopted this legislation yet. The state of New York administers the National Pollutant Discharge Elimination System (**NPDES**). This is a permitting program that is under the State Pollution Elimination Discharge System (SPDES). Any agency that wants to release pollutants from a point source into state waters has to obtain an SPDES permit. There are limitations and strict guidelines that were developed under the Clean Water Act and other limits were set by the New York Department of Environmental Conservation.

The state of Connecticut requires many types of NPDES general permits regardless of whether the release of waste is to a sanitary sewer or other surface waters. For example, NPDES general permits must be obtained for:

- Vehicle washing even if the release is to a sanitary sewer
- Vehicle repair facility floor washwaters
- Parts washwater

The New Jersey Water Pollution Control Act (NJWPCA) fulfills all of the requirements of the federal Clean Water Act (CWA) and requires additional rules that go along with surface and ground water. They also issue general discharge permits to cover dischargers within a geographic boundary (sewer districts, city boundaries and state highway systems). The New Jersey Department of Environmental Protection (DEPE) can also issue a general NJPDES permit to regulate storm water point sources or other point source dischargers.

The Massachusetts Clean Water Act will also apply to surface and ground waters. The act specifically states that no person, public or private group oragencies may dump treated or untreated pollutants into any Massachusetts water without a valid NPDES permit. The permit acts as a legally binding contract between the permit holder and the Commonwealth of Massachusetts. The Massachusetts Environmental Policy Act (MEPA) has made huge changes in the state of Massachusetts. You can visit the site at *www.MEPA.com.* It describes every aspect of MEPA in detail and includes all the regulations and decisions.

Tax Laws [1]

On both the state and federal levels, tax laws are being used to create incentives for environmentally safe products, activities and disincentives against environmentally hurtful products and activities. Some of the approaches that have been implemented or considered are recycling tax credits, taxes on use of virgin materials, taxes on hazardous waste generation, and excise taxes on many products.

Local and Municipal Laws [1]

Most of the environmental regulation is conducted at the state and federal level. Local governments can use their authority to control the location and operation of agencies and facilities in their local jurisdiction. Many of those local issues include:

- Zoning and noise control ordinances
- Nuisance Laws
- Air emission requirements
- Landfill restrictions
- Local emergency planning
- Product recycling incentives

Common Law [1,7]

There is a series of rules and principles that pertain to the government and the security of people and their personal property. These rules and principles have become know as "common law." Originally developed in England these rules were brought to the first American colonies. Following the American Revolution, the rules were adopted and enforced by state and local governments. Common law acquired its authority from basic usage and customs, which are recognized and enforced by the decisions made in a court of law. Civil suits, where a person looks to acquire compensation for some violation of his rights is also common law.

Torts [1,7]

Coming from common law, a "tort" is a private wrong or wrongful act in which an injured party can bring about a civil action. The environmental laws mostly apply to all people including federal agencies, individuals, private organizations, and government contractors. Torts have limited relevance to federal employees. Each person has the responsibility to care for the personal and property right of another person. If any of these rights have been violated, the injured party can receive compensation or restitution from a lawsuit.

Due to the filing of thousands of lawsuits involving asbestos and toxic chemical cases the 1990s have been known as the era of "toxic torts". In the next few sections we will discuss the three most common types of torts that are found in environmental law:
- Nuisance
- Trespass
- Negligence

Nuisance [7]

The definition of a nuisance is "a class of wrongs that are unreasonable or unlawful on his own property, that causes injury or rights of another person or public and produces an inconvenience or discomfort". A person can use his land or personal property as he sees fit with the limitation that the owner of the land uses his property in a reasonable way. When a person uses his property to cause material injury or annoying behavior to his neighbor it is considered a nuisance. It is left up to the courts to determine whether or not the act is considered a nuisance. In order for the act to be considered a nuisance it must cause physical or mental health problems to people under normal conditions.

Noise Nuisance [7] Noise pollution is a nuisance and a common problem everywhere. In order for noise to be a nuisance, it must be so intense where it actually causes discomfort to ordinary people. Noise from industrialization can cause discomfort to people at an extent. There has been a case where homeowners tried to receive compensation and were awarded damages by businesses that were in the neighborhood. Each complaint of noise must be dealt with separately because there are no set standards on the degree or kind of noise nuisances. Other things like smoke, dust, airborne pollutants, water pollutants, and hazardous substances have all been considered nuisances. A nuisance is not an act or failure to act by the person responsible for the condition. It is the condition of the nuisance itself that is the problem.

Coming to a Nuisance[7] If somebody moves into an area where there already is an existing nuisance, he can use a defense known as "coming to a nuisance." When a person moves into a house across from an airport or landfill and then complains of the existing nuisance, a "coming to a nuisance" has happened. Somebody's right to recover damages is not lost because of his prior knowledge of an already existing nuisance. If this person decides to sell the property at the same time of a pending lawsuit, he can still receive compensation.

Trespass [7]
The definition of trespass is the invasion of another person's rights. There must be significant injury to the person, property, or rights of another that has resulted from an unlawful act. A person's intent or motive does not mean anything when determining trespass. Someone can be guilty of trespass even if that person uses reasonable care or good faith. The person's intent or motive affects the level of damages. There are three types of trespass.

  Trespass to personal property: Whether or not physical force was used, it is an injury with the possession of personal property. If you destroy someone's personal property, refuse to return the property, or take property from another they are all offenses of trespass to personal property.

  Trespass to a person: An unlawful act committed against another person. Usually physical contact is involved. Insulting someone or harsh words are not considered trespass.

  Trespass to realty: It is unlawful to force entry on another person's land. One can't interfere with another's personal property or land. The conditions of the land and negligence are irrelevant.

In most pollution cases, the type of trespass is usually trespass to realty. Entry to another person's property can be the result of letting or causing something to cross property boundaries. Throwing material, discharging water, and any wastes invading another person's land are all actions that are considered trespass to realty. Any vibrations or lights that cross property lines are considered nuisances, not trespass.

Negligence [7]

The main points when defining negligence include:

1. The commission or omission of an act;
2. The causing of injury to the plaintiff or his property

The part of the law or tort that deals with acts not intended to cause injury is *negligence*. The case becomes a criminal case if there was intent to inflict injury.

The amount of care and caution that would be taken by an ordinary person in the same situation is defined as the standard of care that is required by law in all negligence cases. A "reasonable man" rule is defined as actions a reasonable person would take under all circumstances. A defendant is liable if his actions are the cause of injury to another.

Under a negligence case, people hurt by the careless and improper handling of hazardous waste can file a suit to seek compensation for their losses. Complying with all governmental regulations and permit conditions is not a defense when negligence has been found to occur. For more information regarding ongoing environmental court decisions go to *www.MassDEPEEnforcementPolicies.com*. At this location you may look up court dates and the outcomes of environmental court decisions.

REFERENCES

1. **Arbuckle, Patton, Blow, and Sullivan**, *Fundamentals of Environmental Law*, Government Institutes, Inc., Rockville, MD, 1993.

2. **Burns, J.A, Peltson, J.W., Cronin, T.E., and Magleby, D.B.**, *Government by the People, Ch. 14, Congress: The People's Branch*, 15th edition, Prentice Hall Publishers, Englewood Cliffs, NJ,1993.

3. **Cozine, M.**, *New Jersey Law Journal*, New Jersey, July 5, Vol.157, 1999.

4. **Gerrard, M.**, *Duty of Consultants, Lawyers to Report Contamination*, New York Law Journal, March 26,Vol.221, 1999.

5. **Silverman, C.**, *The Environmental Lawyer*, New York, N.Y. September 1999.

6. **Umhofer, R., Stitt, K., Savage, A., and Page, R.**, *Environmental Crimes, American Criminal Law Review*, 1999.

7. **Yandel, B. and Meiners, R.**, *Common Law and the Conceit of Modern Environmental Policy*, George Mason Law Review, Vol.7, 1999.

WEB LINKS FOR THIS CHAPTER

Introduction to Laws and Regulations
http://www.epa.gov/epahome/lawintro.htm

EPA Summary of Major Environmental Laws
http://www.epa.gov/Region5/def.html

FindLaw
http://www.findlaw.com/01topics/13environmental

Environmental Law Net
http://www.EnvironmentalLawNet.com

Environmental News Link
http://www.caprep.com/

The Environmental Law Institute
http://www.eli.org

Environmental Law News
http://www.ljx.com/practice/environment/index.html

The Practicing Attorney's Environmental links
http://www.legalethics.com/pa/practice/practice.htm

Pace University School of Law Virtual Environmental Law Library 98
http://joshua.law.pace.edu/env/environ.html

The National Environmental Policy Act of 1969
http://es.epa.gov/oeca/ofa/nepa.html

Massachusetts Environmental Policy Act (MEPA)
http://www.state.ma.us/mepa/301-11tc.htm

Massachusetts DEP Enforcement Policies
http://www.magnet.state.ma.us/dep/enf/enforce.htm#enforce

RESOURCE CONSERVATION AND RECOVERY ACT

2

Travis Veracka and James Bailey

RESOURCE CONSERVATION AND RECOVERY ACT: SUBTITLE C

The Resource Conservation and Recovery Act (RCRA) was enacted in 1976 to manage the large volumes of municipal and industrial solid waste being generated. The RCRA homepage developed by the Environmental Protection Agency (EPA) may be found at *http://www.epa.gov.osw*. RCRA's goals are to protect humans and the environment from waste exposure; to conserve energy and natural resources using recycling and waste recovery techniques; to reduce waste generation; and to ensure that wastes are properly managed. [1]

RCRA is broken down into four distinct categories. Subtitle C establishes a system for controlling hazardous waste from the point of generation to the ultimate disposal, often called the "cradle to grave" system. Subtitle D establishes systems for controlling solid or non-hazardous waste, such as household waste. Subtitle I regulates the underground tank storage of toxic substances and petroleum products *<http://www.epa.gov/epaowser/hotline/ust.htm>* and Subtitle J deals with medical waste *<http://www.epa.gov/epaowser/other/medical/index.htm>*.[2] The framework of hazardous waste management regulations is established under Subtitle C of RCRA. This regulatory framework was designed to protect human health and the environment from improper management of hazardous materials.

RCRA Waste Management Components (40 CFR Parts 261-299)

1. Identification and Listing of Hazardous Waste, 40 CFR 261
2. Standards Applicable to Generators of Hazardous Waste, 40 CFR 262
3. Standards for Treatment, Storage, and Disposal Facilities (TSDs), 40 CFR 264
4. Interim Status Standards for TSDs, 40 CFR 265

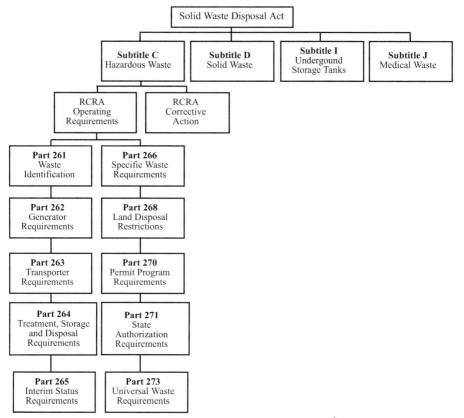

Figure 2.1. Resource Conservation and Recovery Act Tree[3]

5. Interim Status Standards for Owners and Operators of a New Hazardous Waste
   Land Disposal Facility, 40 CFR 267
6. Land Disposal Restrictions, 40 CFR 268
7. Requirements for Authorization of a State Hazardous Waste Program, 40 CFR
   271
8. Standards Applicable to Transporters of Hazardous Waste, 40 CFR 272

Part 261 Waste Identification *<http://www.epa.gov/osw/index.htm>*

Defining what is a hazardous waste is an important question because only those
wastes defined by Congress as hazardous are to be regulated under Subchapter C of
RCRA:

> *solid waste, or a combination of solid wastes,*
> *which because of its quantity, concentration*
> *or physical, chemical or infectious characteristics may:*

(i) cause or significantly contribute to an increase in mortality or an increase in serious irreversible, or incapacitating reversible, illness or; (ii) pose a substantial present or potential hazard to human health or the environment when improperly treated, stored, transported, disposed or otherwise managed.[4]

Using "process knowledge" or laboratory analysis, the generator is responsible for determining if he/she has a hazardous waste. There are two types of hazardous waste, **listed** and **characteristic**. Listed waste includes hundreds of wastes from nonspecific sources (F wastes), manufacturing processes (K wastes), and discarded chemical products (P and U wastes). Characteristic waste is defined as waste which possesses one or more of the following criteria. The four criteria include:

Ignitability. Easily catches fire, with a flash point of less than 140^0 F.

Corrosivity. Easily corrodes materials or human tissue, very acidic of alkaline (pH less than 2.0 or greater than 12.5). Corrodes steel at 6.35mm/yr @ 130^0 F.

Reactivity. Explosive, reacts with water or acid, unstable.

Toxicity. Causes local or systemic damages that result in adverse health effects in an organism. Toxicity determined by the Toxicity Characteristic Leachate Procedure (TCLP).[2]

A solid waste is hazardous if it is named on one of three lists developed by the EPA (Table 2.1):

Listed Hazardous Waste	Characteristic Hazardous Waste	Other Hazardous Waste
Nonspecific sources (F wastes)	Ignitable	Mixtures (hazardous and non-hazardous)
Specific sources (K wastes)	Corrosive	Derived-from wastes (treatment residues)
Commercial chemical products-Acutely hazardous (P wastes)	Reactive	Materials containing listed hazardous materials
Commercial chemical products-Non-acutely hazardous (U wastes)	Toxic	

Table 2.1 RCRA Hazardous Waste Categories

Nonspecific source wastes, F wastes. (40 CFR 261.31) Wastes under this list are generic wastes, commonly produced by manufacturing and industrial processes.

<http://www.epa.gov/epaoswer/non-hw/industd/index.htm>

Specific source wastes, K wastes. (40 CFR 261.32) This list contains wastes from specifically identified industries such as wood preserving, petroleum refining, and organic chemical manufacturing.

Commercial chemical products, P and U wastes. (40 CFR 261.33) This list consists of wastes that are specific commercial chemical products or manufacturing chemical intermediates.[1]

Hazardous Waste Mixtures

A mixture of a listed hazardous waste and a non-hazardous solid waste is considered a hazardous waste. However, if the mixture does not exhibit any hazardous characteristics, then the mixture is deemed non-hazardous.

Derived-from Rule

Any residue from the treatment, storage, or disposal of a listed waste is still a listed hazardous waste unless the waste is recycled to make new products or processed to recover useable material with economic value.[1]

Excluded Wastes

Handlers of these materials are not subject to any hazardous waste regulation.[5]

Industrial ethyl alcohol
Scrap metal
Waste-derived fuels from refining processes
Unrefined waste-derived fuels and oils from petroleum refineries
Petroleum coke

Part 262 Generator Requirements (Table 2.2)

A generator is any person whose act or process produces hazardous waste or whose act first caused the waste to be regulated as a hazardous waste under 40 CFR 260.10.[2]

Generator Classifications *<http://www.gov/epaoswer/hazwaste/gener/index.htm>*

1) Large Quantity Generators

Producer of 1000 kilograms or more of hazardous waste (2200 pounds or more / 265 gallons or more) or more than 1 kilogram of acutely hazardous waste material in a calendar month

2) Small Quantity Generators

Producer of 100 to 1000 kilograms of hazardous waste (220 – 2200 pounds or more / 26 – 265 gallons) and less than 1 kilogram of acutely hazardous waste material in a calendar month

Large and small quantity generators are subject to regulations contained in 40 CFR Part 262. Such regulations include:

EPA ID number
Preparing waste for transportation
Accumulation and storage requirements
Recordkeeping and reporting

Requirement	Conditionally Exempt	Small Quantity	Large Quantity
Amount of waste (40 CFR 261 and 262)	0-100 kg/month	100-1000 kg/month	>1,000 kg/month
Amount of acute waste (40 CFR 261 and 262)	<1 kg	<1 kg	<1 kg
Accumulation (40 CFR 261 and 262)	1000 kg	6000 kg	Unlimited
Storage time (40 CFR 261 and 262)	Unlimited	180 days >200 miles transport then 270 days	90 days
EPA ID number (40 CFR 261 and 262)	N/A	Required	Required
Manifest (40 CFR 262)	N/A	Required	Required
Personnel training	N/A	Some (40 CFR 262.34)	Required
Emergency preparedness equipment and coordination (40 CFR 265 subpart C)	N/A	Basic procedures and a designated emergency coordinator	Full procedures required
Contingency Plan and Emergency Procedure (40 CFR 265 subpart D)	N/A	Basic plan required	Full plan required
Container management (40 CFR 265 subpart D)	N/A	Required	Required
Tank management (40 CFR 265 subpart J)	N/A	SQG requirements	Required
Land Disposal Notice and Waste Management Plan (40 CFR 268.7)	N/A	Required	Required
Hazardous Waste Label	Recommended	Required	Required
Accumulation Date Label	Recommended	Required	Required

Table 2-2. Generator Requirements[3]

3) Conditionally Exempt Generators
Produces no more than 100 kilograms of hazardous waste or no more than 1 kilogram of acutely hazardous waste in a calendar month

Part 263 Transport Requirements
A transporter is defined as any person engaged in the offsite transportation of manifested hazardous waste, by air, rail, highway, or water. A transporter must comply with regulations under 49 CFR Parts 171-179 (The Hazardous Materials Transportation Act) as well as those under 40 CFR Part 263 (Subtitle C of RCRA).[5]

Both state and federal laws impose hazardous material transport requirements. The U.S Department of Transportation is assigned primary responsibility and enforces the Hazardous Materials Transportation Act.

Hazardous waste may be transported from the point of generation to be treated, stored, disposed, or be managed on site. If the waste is to be shipped off site, the generator must prepare a shipping document called a manifest. The manifest must accompany the waste from "cradle to grave." Small and large quantity generators, transporters, and treatment, storage, and disposal facilities (**TSDs**) must obtain an EPA identification number. Form 8700-12 ("Notification of Hazardous Waste Activity") can be referred to for assistance in obtaining an EPA identification number. [2]

A "Uniform Hazardous Waste Manifest" (Form 8700-22) must be used for the transport of RCRA regulated hazardous waste. Supplemental state manifest may be used if adapted from the EPA manifest. Waste manifest require the following information:
1. Names, addresses, telephone numbers, and EPA identification numbers of the generator, transporter, and the designated recipient facility
2. Descriptions of each unit of hazardous waste in the shipment, including U.S. Department of Transportation (DOT) description, EPA waste code number, type and quantity of all containers, and any additional description
3. Special handling instructions and additional information
4. Land disposal restriction requirements (LDRs)
5. Signed certification by the generator attesting to the truth of the information, and, for LQGs, of the existence of a waste minimization program

Manifests
Hazardous waste that is sent to a TSDF or an approved designated facility must be listed on a Uniform Hazardous Waste Manifest
<*http://epa.gov/epaoswer/hazwaste/gener/manifest/index.htm*>. A manifest provides a description of the hazards of the waste and the waste handlers. You must sign and date the manifest and obtain the signature of the transporter on the manifest. You must also keep a copy of the form until you receive a copy signed by the TSDF. Manifests must be kept for at least 3 years from the date of shipment.[5]

Land Disposal Restriction
Many types of hazardous waste are restricted from being disposed in or on the land

due to the probability of groundwater or soil contamination. The manifest for these wastes must be accompanied by a land disposal restriction (LDR) notification. The LDR is a one-time notification form that is signed by you, the generator, and indicates that you understand that this waste cannot be land disposed. According to federal regulations, it is the responsibility of the generator to provide the manifest and the LDR. Most hazardous waste disposal contractors will supply one or both of these forms for you. [5]

Exception Report

Once a hazardous waste manifest has been signed and dated by you (generator), EPA gives the disposal contractor 35 days to transport the waste to the TSDF and return a signed copy of the manifest. If, after 35 days, you do not receive the "Return to Generator" copy of the manifest signed by the TSDF, you must notify the transport and/or the TSDF to determine the status of the hazardous waste. If, after 45 days, the "Return to Generator" copy is not received, you must file an exception report. The exception report must include a legible copy of the manifest and a cover letter signed by you explaining efforts taken to locate the hazardous waste and the results of those efforts. This report must be maintained for at least 3 years. [5]

Biennial Report

Large quantity generators of hazardous waste must submit a biennial report by March 1 of each even numbered year (i.e., 1990, 1992, etc.) on EPA Form 8700-13A. The report basically covers the amount and types of hazardous waste generated in the previous two years and the TSDF receiving the waste. This report must be maintained for at least 3 years. [5]

Miscellaneous

Generators are also required to keep records of test results, waste analyses, or other waste determinations for at least 3 years from the date of shipment of that waste.

Storage/Containers

You can store hazardous waste in tanks, containers, drip pads, or containment buildings (40 CFR Part 262.34). Tanks and containers designated for storage must not be leaking, bulging, rusted, or incompatible with the waste stored in them (e.g., certain types of acid in metal containers). Storage areas must have secondary containment, an alarm, a fire extinguisher, a "No Smoking" sign, and a means of communication (e.g., walkie-talkie, cellular phone, or air horn) in the event of a spill or other emergency. [5]

Labeling

Hazardous waste containers must be labeled with the words "Hazardous Waste," the contents of the container, the accumulation start date, the waste codes of the contents, and the EPA ID number of the generator. A standard yellow hazardous waste label can usually be obtained through your EPA or State office, or through some catalogs. [2]

Storage/Time Limitations

LQGs are required to dispose their hazardous waste within 90 days of placing the waste into the container. The exception to this is if the facility elects to "satellite" its waste first. If you satellite your waste, you must collect it in a container "at or near the point of generation." The interpretation of "at or near" varies from state to state but generally means the container cannot be separated by a door or wall from the point of generation and cannot be more than 50 feet from the point of generation. The container must be labeled only with its contents. Most states allow satellite generation of up to 55 gallons of one type of waste. Once the 55-gallon limit has been reached, the waste must be moved to an onsite accumulation area or building where it can be stored for up to 90 days from that time.[1]

Requirements for Transporters

1. Obtaining an EPA identification number
2. Handling waste properly before shipment (packaging, labeling, marking, placarding, accumulation time, etc.)
3. Complying with the manifest system
4. Recordkeeping and reporting requirements

When a serious accident or spill occurs, the transporter must notify the National Response Center (NRC) at 1-800-424-8802. The NRC must be notified when:

1. A person is killed or seriously injured
2. Estimated damage exceeds $50,000
3. The spill involves disease-causing agents or radioactive material
4. The spill exceeds a Superfund reportable quantity
5. A life-threatening situation exists.(EPAOM) [5]

Part 264 Treatment, Storage, and Disposal Requirements

Subtitle C requires all treatment, storage, and disposal facilities (TSDFs) that handle hazardous wastes to obtain an operating permit and abide by treatment, storage, and disposal (TSD) regulations *<http://www.epa.gov/epaoswer/osw/laws-reg.htm>*. [5]

According to 40 CFR 260.10, a TSDF encompasses three different functions:

Treatment – Any method, technique, or process, including neutralization, designed to change the physical, chemical, or biological character or composition of any hazardous waste so as to neutralize it, or render it non-hazardous or less hazardous, or to recover it, or render it safer to transport, store or dispose of, or amendable for recovery, storage, or volume reduction

Storage – The holding of hazardous waste for a temporary period, at the end of which the hazardous material is treated, disposed of, and stored elsewhere.

Disposal – The discharge, deposit, injection, dumping, spilling, leaking, or placing of any solid into or on any land or water so that the waste or any constituent thereof may enter the environment or be emitted into the air or discharged into any waters, including ground waters.

Subpart A – Who is subject to regulation?

In general, all owners or operators of facilities treating, storing, or disposing of hazardous waste must meet the appropriate TSD regulations. The exceptions include:

1. Farm disposing of self used pesticides
2. The owner and operator of a totally enclosed treatment facility
3. The owner or operator of an elementary neutralization unit or a wastewater treatment unit
4. A person cleaning up a hazardous spill or discharge
5. Facilities that reuse, recycle, or reclaim hazardous wastes
6. Generators accumulating wastes within time periods specified in 40 CFR Part 262
7. A transporter storing manifest shipments less than 10 days
8. A facility regulated by an authorized state program

Subpart B – General Facility Standards

Before a facility owner or operator can handle hazardous waste, they must apply for an EPA identification number. Owners and operators must also ensure that their wastes are properly identified and handled. Their facilities must be secure and operating properly. Their personnel must also be properly trained in hazardous waste management. To satisfy these conditions, owners and operators must take the following actions:

Conduct Waste Analysis – This ensures that owners and operators possess sufficient information on the properties of the waste they manage to be able to treat, store, or dispose of them in a manner that will not pose a threat to human health or the environment.

Install Security Measures – The security requirements prevent unknowing entry of people and minimize the potential for the unauthorized entry of people or animals onto the active portions of the facility.

Conduct Inspections – Owners or operators must develop an inspection schedule to assess the status of the facility and detect possible problem areas. Inspection records must be kept on file for three years.

Conduct Training – This is accomplished by ensuring that facility personnel acquire expertise in the areas to which they have been assigned.

Properly Manage Ignitable, Reactive, or Incompatible Wastes – All ignitable or reactive wastes must be protected from sources of ignition or reaction or treated to remove the cause for concern.

Comply With Location Standards – Location standards prohibit siting a new facility in a location where flood or seismic events could affect a waste management unit. Bulk liquid wastes are also prohibited from placement in salt domes, salt beds, or underground mines or caves.

Subparts C and D – Preparedness, Prevention, Contingency Plans, and Emergency Procedures

The preparedness and prevention requirements are explicit (e.g., installing fire protection equipment and alarms and arranging for coordination with the local authorities in emergency situations). Contingency plan requirements require an owner or operator to develop an action plan for emergency procedures.

Subpart E – Manifest System, Recordkeeping, and Reporting

Requirements under this section create a manifest system that specifies that the manifest must be returned from the facility owner to the generator. The manifest system is established in the manifest regulations 40 CFR Part 262.

When the owner of the TSDF receives the waste, it must be ensured that the waste in the truck is the same as indicated on the manifest. This ensures that there are no discrepancies in the amount of type of hazardous waste that was sent by the generator. If there is a discrepancy, it must be reconciled with the generator. If this is impossible, the EPA must be notified within 15 days of the incident. Subpart E also includes requirements for recordkeeping and reporting. Records include operating records, biennial reports, unmanifested waste reports, and reports on releases, groundwater monitoring, and closure.

Subpart F – Standards for Hazardous Waste Treatment, Storage, and Disposal Units *<http://www.epa.gov/osw/index.htm>*

To reduce the potential for lechate to threaten environmental or human health, EPA has developed design and operating standards to detect, contain, and clean up any possible leaks. Areas of development include the following:

Containers
Containment buildings
Drip pads
Land treatment units
Landfills and waste piles
Surface impoundment
Tanks

Subpart G – Closure

When TSDFs stop receiving waste for treatment, storage, or disposal the owner and operator must either remove the waste or leave the waste in place where it poses no future threat to human health and to the environment. The closure and post-closure regulations consist of two parts, the general provisions and the technical standards for specific types of hazardous waste management units.

Closure is the period directly after the TSDF stops its normal operations. Post-closure is after the closure period at which time the owners and operators conduct monitoring and maintenance activities to maintain the disposal system and prevents a possible release.

Subpart H – Financial Assurance

To prevent a facility from stopping operations and failing to provide for the closure and post-closure of the facility, EPA ensures that the facility owners have demonstrated that they have the financial resources to conduct the proper closure and post-closure of the facility.

Subpart I – Ground Water Monitoring

RCRA requires TSDF owners and operators of land-based treatment, storage, or disposal units (landfills, surface impoundment, and land treatment units) to monitor the groundwater under their facility.

Subpart J – Air Emission Standards

Owners and operators of TSDFs that produce volatile organics with an average annual total of 10 ppm must reduce organic emissions at their entire facility. The owner must either modify the procedures used or install devices to control organic emissions in order to meet the standards.

Part 265 Interim Status Requirements

Interim Status is for facilities or entities that have gone into business but have not yet obtained a RCRA permit for operation. Interim Status also applies to entities that have lost their RCRA permit for operation.[5]

The objective of the interim status technical requirements is to reduce the potential for environmental and public health exposures resulting from hazardous waste treatment, storage, and disposal. An owner or operator of an interim facility can find the applicable requirements in Subparts F through R of 40 CFR Part 265. Interim status requirements are divided into two groups:

1. General standards applying to several types of facilities
2. Specific standards applying to each waste management method

Part 266 Special Waste Requirements

This part gives special requirements for recyclable materials used in a manner constituting disposal, recyclable materials utilized for precious metal recovery, and hazardous waste burned in boilers and industrial furnaces.

Part 268 Land Disposal Restrictions

40 CFR Part 268 gives requirements and restrictions for land disposal of hazardous waste. Areas that are covered include dilution prohibition, schedule for land disposal prohibition and establishment of treatment standards, ban on land disposal of certain specific wastes, treatment standards, prohibitions on storage, and exemptions from requirements.

The LDR program was designed to protect ground water from contamination by requiring hazardous waste to be physically or chemically altered to reduce the toxicity or mobility of waste constituents prior to disposal
<http://www.epa.gov/epaoswer/hazwaste/ldr/index.htm>.[1]
Waste must be a RCRA waste in order to be regulated under the LDR program. The LDR program consists of three components:

> Disposal prohibition
> Dilution prohibition
> Storage prohibition

Disposal Prohibition

The disposal prohibits the land disposal of hazardous waste that has not been adequately treated. Treatment standards can be either concentration levels or treatment technologies that must be performed on the hazardous waste material. The treatment standards are based on Best Demonstrative Available Technology (**BDAT**). [5]

Both listed and characteristic wastes must meet LDR treatment standards before they can be disposed. Wastes are also examined for underlying hazardous constituents. The underlying hazardous constituents must be treated so that they meet contaminant-specific levels that are referred to as universal treatment standards (UTS). UTS are listed in 40 CFR 268.48.

Dilution Prohibition

Dilution prohibition prevents the act of attempting to reduce the hazard of a waste by dilution. The dilution prohibition states that it is impermissible to dilute hazardous waste to circumvent proper disposal.[5]

Storage Prohibition

EPA regulations state that if waste storage exceeds one year, the facility must ensure that the waste is being properly maintained in order to accumulate quantities necessary for effective treatment or disposal. Waste storage for less than one year directs the EPA to assume the responsibility to prove that such storage is not the purpose of accumulating quantities necessary for effective treatment or disposal.[5]

Part 270 Permit Requirements

Permits are used to identify the standards that facilities must adhere. The EPA or authorized stated can issue permits <*http://epa.gov/epaoswer/hazwaste/ permit/index.htm*>. Permits that are required include permit application, permit conditions, changes to permits, expiration and retention of permits, and qualifications for and termination of interim status.

Treatment, Disposal, or Storage without a Permit Circumstances

1. Generators storing waste on site for less than 90 days
2. Small quantity generators who store waste on site less than 180 days
3. Farmers disposing of their own waste pesticides on site
4. Owners operating of totally enclosed treatment facilities, waste water treatment units, and elementary neutralization units
5. Transportation storing manifested wastes at a transfer facility for less than 10 days
6. Persons engaged in containment activities during an emergency response
7. Owners or operators of solid waste disposal facilities handling only conditionally exempt small quantity generator waste
8. Person engaged in Superfund on-site cleanups and RCRA Section 7003 cleanups.[1]

Types of Hazardous Waste Permits

Treatment, Storage, and Disposal Permits – The units that are included are: containers, tank systems, surface impoundment, waste piles, land treatment units, landfills, incinerators, and miscellaneous units.

Research, Development, and Demonstration Permits – Alternative treatment techniques are encouraged by issuing research, development, and demonstrative (RD and D) permits for innovative and experimental treatment technologies.

Post-Closure Permits – Land disposal facilities must obtain a post closure permit if they leave waste in place when they close the facility.

Emergency Permits – Where there is an "imminent and substantial endangerment to human health and the environment", the EPA can forego the normal permitting process. A temporary (90-day or less) permit can be issued:

1. Non-permitted facility for the treatment, storage, or disposal of hazardous waste
2. Permitted facility for the treatment, storage, or disposal of hazardous waste

Permit-by-Rule – A permit-by-rule eliminates the need for waste facilities to submit a full subtitle C permit application. Facilities may fall under this category if they are permitted under the following:

1. Safe Drinking Water Act (Underground Injection Control permit)
2. Clean Water Act (National Pollutant Discharge Elimination System permit)
3. Marine Protection, Research, and Sanctuaries Act (Ocean Dumping permit)

Trial Burn and Land Treatment Demonstrative Permits – Permits are issued to construct and operate new hazardous waste management facilities. Land treatment facilities and incinerators must undergo a trial period in order to test the efficiency of their system. This period is called a trial burn for incinerators and a land treatment demonstration for land treatment facilities.[5]

Permitting Process

All hazardous waste TSDFs require getting a RCRA permit following the same basic permitting process.

1. Submitting a permit application
2. Reviewing the permit application
3. Preparing the draft permit
4. Taking public comment
5. Finalizing the permit [1]

Permit Administration

RCRA permits are valid for up to ten years. Land disposal permits have an additional requirement in that they must be reviewed after five years from the issued date. During the term of the permit its status could change. Situations that may arise concerning the permit include the following:

1. Modified
2. Revoked, and reissued
3. Terminated [1]

Corrective Action Process

Congress expanded EPA's authority through HSWA to address releases of hazardous materials through corrective action. Corrective action requirements are imposed through a permit or an enforcement order. RCRA facilities with permits must, at a minimum, contain schedules of compliance to include releases and provisions of financial assurance to cover the cost of possible corrective measures.[5]

Corrective Action Components

RCRA Facility Assessment (RFA)
RCRA Facility Investigation (RFI)

Corrective Measures Study (CMS)
Corrective Measures Implementation (CMI)[5]

The Environmental Priorities Initiative (EPI)

The EPI is a joint RCRA/Superfund screening approach used to prioritize the environmental significance of facility cleanup sites. The ranking of these sites is based on the threat that each site possesses to human health and the environment.[1]

Part 271 State Authorization Requirements

The ultimate intention for RCRA was for the states to assume primary responsibility for implementing hazardous waste regulations, with oversight from the federal government. For a state to take responsibility for its hazardous waste management, it first must be authorized to do so by the EPA. Through state authorization, the EPA sets minimum federal standards to prevent overlapping state regulatory programs. States that have been approved for self- hazardous waste control are known as authorized states <*http://epa.gov/epaoswer/hazwaste/authmain.htm*>.[5]

Under RCRA, states have two options for assuming waste responsibility, final or interim authorization.

Final Authorization

A state must be fully equivalent to, no less stringent than, and consistent with the federal program. States may however implement hazardous waste management programs that are more stringent than federal requirements.

Any state that attempts to obtain final authorization for its hazardous waste program must submit an application to the EPA administrator containing the following elements:

1. A letter from the governor requesting program authorization
2. A complete description of the state hazardous waste program
3. An attorney general's statement
4. A memorandum of agreement (MOA)
5. Copies of all applicable state statutes and regulations
6. Documentation of public participation activities

Interim Authorization

States that do not meet the minimum federal requirements for hazardous waste regulations can obtain interim authorization. Interim authorization is a temporary method of promoting state involvement in state hazardous waste management while encouraging states to develop a program that is equivalent to the federal program.

Part 273 Universal Waste Requirements

Regulated under these requirements are batteries, pesticides, thermostats, and household and conditionally exempt small quantity generators. The purpose of this rule is to reduce the amount of waste that is entering the municipal waste stream. The

Universal Waste Rule encourages recycling and proper disposal of certain common hazardous waste items.[3]

1. Batteries
2. Agricultural Pesticides
3. Thermostats
4. Lamps

There are four regulated participants in the universal waste system: small quantity handlers of universal waste, large quantity handlers of universal waste, universal waste transporters, and universal waste facilities.[6]

Small Quantity Handlers of Universal Waste

1. Accumulate less than 5000 kilograms (11,000 pounds) of all universal waste categories combined at their location at one time
2. Accumulation time is one year

Large Quantity Handlers of Universal Waste

1. Accumulate more than 5000 kilograms (11,000 pounds) of all universal waste categories combined at their location at one time
2. Accumulation time is one year
3. Maintain shipping records
4. Obtain EPA ID number
5. Employee training

Universal Waste Transporters

1. Do not need a RCRA waste manifest but must comply with DOT transport regulations
2. May store waste for up to 10 days at a transfer facility

WASTE MINIMIZATION:
THE KEY TO BETTER WASTE MANAGEMENT

The easiest and most cost-effective way of managing any waste is not to generate it in the first place. You can decrease the amount of hazardous waste your business produces by developing a few "good housekeeping" habits. Good housekeeping procedures generally save businesses money, and they prevent accidents and waste. To help reduce the amount of waste you generate, try the following practices at your business.

Do Not Mix Wastes

Do not mix non-hazardous waste with hazardous waste. Once you mix non-hazardous waste with hazardous waste, you may increase the amount of hazardous waste created, as the whole batch may become hazardous. Mixing waste can also make recycling very difficult, if not impossible. A typical example of mixing wastes would be putting non-hazardous cleaning agents in a container of used hazardous solvents.

Recycle and Reuse Manufacturing Materials

Many companies routinely put useful components back into productive use rather than disposing of them. Items such as oil, solvents, acids, and metals are commonly recycled and used again. In addition, some companies have taken waste minimization actions such as using fewer solvents to do the same job, using solvents that are less toxic, or switching to a detergent solution.

Change Materials, Processes, or Both

Businesses can save money and increase efficiency by replacing a material or a process with another that produces less waste. For example, you could use plastic blast media for paint stripping of metal parts rather than conventional solvent stripping.

Safely Store Hazardous Products and Containers

You can avoid creating more hazardous waste by preventing spills or leaks. Store hazardous product and waste containers in secure areas, and inspect them frequently for leaks. When leaks or spills occur, materials used to clean them up also become hazardous wastes.[7]

REFERENCES

1. **EPA**, *RCRA Orientation Manual,* Office of Solid Waste, Washington, D.C. 20460, EPA/530-SW-90-036, 1990.

2. **Moore, G.,** *Environmental Regulations and Compliance*, School of Public Health, University of Massachusetts, 1997.

3. **EPA,** *RCRA Requirements,* Office of Regulatory Enforcement Division, *http://es.epa.gov/oeca/ore/rcra/rcraregfrmwrkb.html*, 2000.

4. **EPA**, *RCRA Vs Code Annotated,* Title 42, pt. 6901 et seq. 1995.

5. **EPA**, *RCRA Orientation Manual,* Office of Solid Waste, Washington, DC 20460 EPA/530-SW-98-036,1998.

6. **EPA**, *Managing Hazardous Waste – RCRA Subtitle C,* Office of Solid Waste, 2000.

7. **EPA,** *EPA Understanding Hazardous Waste Rules – A Handbook for Small Businesses,* EPA ID Number. EPA 530-k-95-001, 1996.

8. **EPA,** Welcome to RCRA Online, EPA Office of Solid Waste, http://yosemite.epa.gov/osw/rcra.nsf/, 2000.

WEB LINKS FOR THIS CHAPTER

Authorized States
http://epa.gov/epaoswer/hazwaste/authmain.htm

EPA Catalog of Solid and Hazardous Waste Publications
http://www.epa.gov/epaoswer/osw/catalog.htm

EPA RCRA HomePage
http://www.epa.gov/osw/index.htm

EPA Resource Conservation and Recovery Information System
http://www.epa.gov/enviro/html/rcris/rcris_overview.html

Full Text of RCRA
http://www4.law.cornell.edu/uscode/unframed/42/ch82.html

Generator Classifications
http://www.gov/epaoswer/hazwaste/gener/index.htm

Massachusetts Licensed Site Professional Homepage
http://www.magnet.state.ma.us/lsp/lsphome.htm

Massachusetts Hazardous Waste Transportation
http://www.magnet.state.ma.us/dep/matrix.htm

Medical Waste
http://www.epa.gov/epaowser/other/medical/index.htm

Permits
http://epa.gov/epaoswer/hazwaste/permit/index.htm

RCRA Summary
http://www.epa.gov/reg5oopa/defs/html/rcra.htm

Treatment, Storage, and Disposal (TSD) Regulations
http://www.epaoswer/osw/laws-reg.htm

Underground Tank Storage of Toxic Substances and Petroleum Products
http://www.epa.gov/epaowser/hotline/ust.htm

Uniform Hazardous Waste Manifest
http://epa.gov/epaoswer/hazwaste/gener/manifest/index.htm

CERCLA

3

Jeff Bagg

THE COMPREHENSIVE ENVIRONMENTAl, RESPONSE, COMPENSATION, AND LIABILITY ACT (CERCLA)

Ever since the Industrial Revolution, the U.S. has been inundated with technology. This industrial technology has been a key component in the formation of this nation. Benefits cannot occur without costs, though the costs are increasing in size and severity. Many of our adaptations as humans come with great environmental expenses. Our agricultural techniques rely heavily on things such as pesticides and insecticides which are known to percolate into underground water tables. The creation and manufacturing of plastics was a major step in human progression, but it hasn't come without a cost. Producing plastics leaves volatile organic compounds (VOCs), chemical additives, cadmium and polymers at abandoned industrial sites. It is well documented that both active and inactive paper mills have long polluted environments with chlorinated compounds such as dioxin. [1]

The rate at which technology has progressed is phenomenal. The invention of the gasoline engine in the early twentieth century was both critically important and dangerous. The production, use, and need for petroleum has never been greater, and the environmental costs of that are still not fully known. The movement into the twenty first century places substantial emphasis on the wastes being created by humans as a race. Every human produces wastes and has an environmental impact, some cultures more than others. The fact is that this waste must go somewhere. Here, in the U.S., the majority makes its way to landfills. The content of landfills is often known to contain both benign and dangerous materials. Some chemicals commonly found in landfills could range from household products and cleaners, methane and pesticides, to polychlorinated biphenyl (PCBs).[1] The average person has probably seen or been to a landfill. Landfills are usually located on the outskirts of urban areas and are filled with our waste products. These landfills become heaping piles of refuse and are rarely given second thoughts. We must begin to consider that every thing consumed had to be produced. We are both a consumer and producer society, every thing is composed of certain elements. The problem is that each finished product also creates wastes. These wastes are often unsafe and hazardous compounds. Similar to consumers, industries must also dispose of their byproducts, and this they have done.

Up until the mid/late 1970s industries had no guidelines or regulations for the disposal of hazardous wastes. Decades of activities such as spilling, discharging, leaching, dumping, and burying created and abundance of hazardous waste sites nationwide. [2] It wasn't until the discovery of an enormous abandoned toxic waste site called Love Canal that this problem was given national attention. Love Canal in Niagra Falls, New York, became infamous in the late 1970s when years of illness, odors, and seepage culminated into a revelation exposing this massive toxic waste dump. From the 1950s through most to the 1970s, residents had noticed strange odors in the air and unusual seepage into their basements and yards. Abnormally high incidents of reproductive disorders and urinary tract disorders were being noticed and were finally cataloged by Lois Gibbs, the areas first and foremost activist. [3]

The combination of media coverage, intense environmental lobbying, and a growing tide of public pressure created a need for a congressional response. Congress reacted by creating the Comprehensive Environmental Response, Compensation and Liability Act (CERCLA) on December 11, 1980. CERCLA was instituted to address the cleanup and remediation of inactive hazardous waste sites and to limit the threat to human health and the environment posed by these contaminated sites.[4] CERCLA, also known as Superfund, was initially created to clean up the nation's worst abandoned hazardous and toxic wast sites. Over a five year span a trust fund of $1.6 billion was allotted using the revenue raised from taxes placed on chemical and petroleum industries along with environmental taxes on corporations. [5]

Superfund at a Glance

Superfund is managed by the Environmental Protection Agency (EPA), in collaboration with each state's environmental agency. In Massachusetts it is the Department of Environmental Protection (DEP)
<*www.state.ma.us/dep/dephome.htm*>. Through Superfund, the EPA has the authority and ability to locate, investigate, and clean up hazardous waste sites. The Superfund program gives the EPA the capacity to

1. Establish prohibitions and requirements relating to closed and abandoned sites.
2. The EPA is given a provision that allows for the use of strict, joint, and several liability. This enables the EPA to fine any industry, active or inactive, even industries which acted lawfully in the past.
3. CERCLA/Superfund has access to a trust fund generated by a tax imposed on chemical and petroleum industries to compensate for the expenses of a cleanup when a responsible party cannot be identified.[3]

What Is a Hazardous Waste Under CERCLA?

The method the EPA has derived to define a hazardous waste considers chemicals which are listed under other hazardous waste and pollution laws. The laws which contain these chemicals are the Clean Water Act (**CWA**), the Resource Conservation and Recovery Act (**RCRA**), and the Toxic Substance Control Act (**TSCA**). The Clean Water Act originally established in 1948, separates water pollution into two broad classifications. The first consists of point sources. These are sources which can be identified as a specific waste stream that can be tracked and regulated. The second are non-point sources. These contaminants are usually in the form of runoff from places like roofs, roads, and other areas exposed to rain. Because these sources are either difficult to trace or unknown, it is difficult to regulate and monitor them. Under RCRA, chemical compounds fall under two categories, listed and characteristic wastes. A listed waste can originate from many places, including non-specific sources, manufacturing processes, and discarded chemical products. A detailed list of these hazardous wastes can be found in the Code of Federal Regulations, title 40, part 261, subpart D.[4] The characteristic wastes fall under four criteria which evaluate the potential risks. The first is ignitability. This group represents chemicals which can catch fire at temperatures less that 140 degrees. The second criteria is corrosivity. These substances can easily corrode materials and can severely damage human tissue. These substances usually contain a pH less than 2 or greater than 12.5. The third group represents the level of reactivity of a compound. The compounds are typically explosive in nature, react with both water and acids, and for the most part are unstable. The fourth and last category under RCRA is toxicity. These are materials which can, and often due, cause adverse health effects in organisms. The last group of chemicals considered hazardous under CERCLA are contained within TSCA. Chemicals considered imminently hazardous and pose unreasonable risk to human health are regulated here. TSCA requires the creation of data which details the effects of all manufactured chemicals on human health and the environment. TSCA also has the ability to regulate, restrict, and prohibit the manufacture and use of certain chemical compounds such as asbestos, radon, PCBs and CFCs. Some of the chemicals which are not regulated under CERCLA are oil, petroleum, and natural gas. CERCLA does however confront the problems caused by oil spills such as cleanup, costs, and prevention through the National Oil and Hazardous Substances Pollution Contingency plan. [4]

Identifying the Source

Now hopefully we have established some background information which is crucial to reveal the possibilities for success within the Superfund program. In order for the EPA to become involved in reclaiming a hazardous site, they must first be made aware of it. Historically, this has happened in a variety of ways and for different reasons. At the local and state levels, agencies and businesses often unearth these areas of contamination. On the Federal level, there is the EPA, which is the leading source for the discovery of hazardous waste sites.[5] More often than not though, it is the citizens in areas closest to the problem who demand the attention of the authorities. One example of this can be the General Electric plant in Pittsfield, Massachusetts. It is here that a very important debate over Superfund is ongoing. The city of Pittsfield is plagued with wide-spread PCB contamination. It began as the city was enjoying a post war boom in the 1940s, and the industrial giant GE offered the use of its debris as fill for anyone who wanted it. For decades, the fill went out in truckloads, to backyards and commercial sites including the filling and redirecting of 11 natural oxbows along the Housatonic River. [6] It was around 1968 that the contamination began to be uncovered. A storage tank located in a building on the GE property collapsed, releasing a liquid PCB (Aroclor 1260) into the riverbank soil and the Housatonic's riverbed sediments. Ever since then local citizens have come to the realizations that their homes may be sitting atop a toxic waste dump. One property owner, interviewed in a local newspaper, tells how his property is surrounded by chain-linked fences showing signs warning of contamination. According to an excerpt from the newspaper article, a study funded by GE in collaboration with researchers from the National Cancer Institute shows no apparent links between PCB's and cancer, but local citizens tell a different story. Many of the towns older residents have seen friends and family pass away because of cancer. " The cancer rate is real here", a gentlemen shares his fears with a quote in the paper. "Certain trends make me nervous, my grandmother died of cancer, my mother died of cancer, my uncle died of cancer." [6] The communities in Berkshire county concerned with GE and PCB contamination have a Web site full of information, it is *<www.housatonic-river.com>* and gives a very comprehensive and personal account of the entire situation. These types of scenarios really show the importance and effectiveness of citizen involvement.

What is a Release?

Under CERCLA, the term release is defined broadly as to include virtually any situation leading to a hazardous substance accidentally being freed from its normal container. Therefore a release can occur whenever "any spilling, leaking, pumping, pouring, emitting, emptying, discharging, injecting, escaping, leaching, dumping, or disposing occurs within an environment."[2] Federal facilities must also follow certain guidelines with respect to hazardous materials. To assure federal compliance with the laws under CERCLA, all agencies must complete a hazardous waste docket. These dockets are used to keep track of federal facilities which manage, produce, or may potentially release hazardous substances. The facilities are required to notify the EPA of any potential, known, or suspected releases. If a facility is placed on the docket, the industry or agency may then be required to complete expensive, comprehensive studies and possible remedial actions.[2] If you wish to find out if a facility near you handles hazardous waste, the EPA has created a query page which can be found at *<www.epa.gov/enviro/html/rcris_query_java.html>*. This search requires either the name, location, and/or industrial classification in order to locate and find information about its environmental records.

Response Authorities

It was in 1968 that the first comprehensive system for accident reporting, spill containment, and site cleanup was created. This established a response headquarters and national and regional reaction teams. This plan acted as the federal government's blueprint for responding to both oil spills and hazardous waste releases. This coordinated response is known as the National Contingency Plan (**NCP**) and was developed in response to a massive oil spill from the *Torrey Canyon* off the coast of England in 1967. More than 37 million gallons of crude spilled into the water and due to the lack of response caused devastating environmental damages. [8] Since its creation in 1968, the NCP has been revised several times. After the Clean Water Act (CWA) of 1972, the framework for responding to hazardous substance spills as well as oil discharges were created. When Superfund legislation *<http://www.epa. gov/superfund/whatissf/cercla.htm>* was enacted in 1980, the NCP was broadened to cover releases at hazardous waste sites requiring emergency removal actions. The most recent revision occurred in 1994 in response to the Oil Pollution Act of 1990 *<http://www.epa.gov/oilspill/opaover.htm>*. [8] The revisions within the NCP concerning hazardous substance response plans have established procedures and standards for responding to releases of hazardous substances, pollutants, and contaminants. It has also created methods for discovering and investigating contaminated areas.

Techniques have been implemented to evaluate both the possible remedies and costs of reducing the danger to public health. The EPA has constructed criteria for determining the extent of either removal or remediation of a site. It considers risks to populations, potential drinking water contamination and possible destruction of sensitive environments. The ability to designate roles and responsibilities among federal, state and local governments, along with the means necessary to assure cost-effective remedial actions have been created through the NCP. To examine the original National Contingency Plan, it can be found in Section 1321 of Title 33 at *<http://www4.law.cornell.edu/uscode/unframed/33/1321.htm>*. [9]

Environmental laws can clearly be influenced and changed in response to certain environmental tragedies, CERCLA is no exception. In 1988, Congress passed Title III of the Superfund Amendments and Re-authorization Act (SARA), also referred to as the Emergency Planning and Community Right-to-Know Act (EPCRA), in response to a devastating chemical accident in Bhopal, India which killed thousands. EPCRA was the culmination of a series of social and political events that focused attention on the potential for chemical accidents and their impacts on humans and the environment. Historically, chemical use has preceded and outstripped any knowledge of the impacts on human health, the environment, and the community. In response to the public's growing concern over the safety of operations and materials at industrial facilities, two goals were attempted with SARA III. First, to facilitate and promote planning for chemical emergencies at state and local levels. Secondly, to provide information to the public about the chemicals used, stored, and released into local communities.[10]

The Ranking and Listing of Hazardous Sites

The method for evaluating contaminated sites is the Hazardous Ranking System **(HRS)**. It is a numerically based screening system that uses information from initial, limited investigations. These techniques are known as preliminary assessments and site inspections. They assess the relative potential of a site to be a threat to human health or the environment. Preliminary assessments are used to distinguish between sites that pose little threat and sites that may require emergency response actions. The site inspection focuses attention on collecting samples to determine which substances are present and if they are being released in nearby environments. Another step in the ranking process is using detailed studies called remedial investigation/feasible study **(RI/FS)**. These studies are used to conduct treatability tests to evaluate the potential performance and cost of the proposed treatment technologies. Within this system four major pathways are examined. The pathways include drinking water, surface water, soil exposure, and air migration.[11]

The National Priorities List

After the information from the Hazardous Ranking System is gathered and evaluated, the EPA uses it as a guide to determine whether the site will require further investigation. This knowledge enables them to assess the nature and extent of the human health and environmental risks of a site. The NPL is also used to notify the public of sites which the EPA believes may pose health risks and need further investigation. The NPL functions primarily as an informational strategy to identify for the states and the public which sites may need remedial actions.[12]   In order for a site to have long term remedial actions payed for by the Superfund, they must be placed on the list. The effectiveness of the NPL has become better and better and has allowed Superfund to move and act more efficiently. In the last six years, the total number of NPL sites that have met all cleanup goals for at least one area of contamination has tripled.[13] The following charts, found on the EPA's superfund page *<http://www.epa.gov./superfund/whatissf/mgmtrpt.htm>*, represent the fiscal years of 1993 (Fig. 3-1) and 1997 (Fig. 3-2). The emphasis should be placed on the percent increase of completed construction within the two periods.[14]

If you are interested in finding out whether your state has any facilities near you listed on the NPL, you can view maps of sites within your area at *<http://www.epa.gov/superfund/sites/npl/npl.htm>*.

When the Cleanup Begins

When it comes to paying for a Superfund cleanup, the EPA has two options under CERCLA. The EPA can conduct the cleanup and then seek cost recovery from pollut-

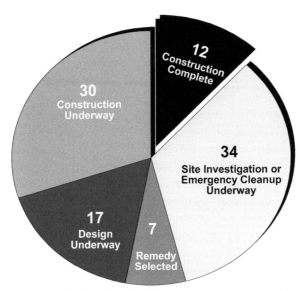

Figure 3.1. Progress and Steps in the Cleanup of Superfund Sites, January 1993.
Adapted from *<http://www.epa.gov/superfund/whatissf/mgmtrpt.htm>*

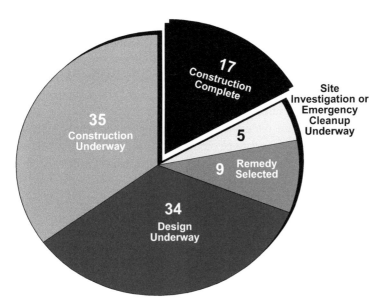

Figure 3.2. Progress and Steps in the Cleanup of Superfund Sites, January 1997.
Adapted from *<http://www.epa.gov/superfund/whatissf/mgmtrpt.htm>*

ing industries, known as Potentially Responsible Parties (**PRPs**), which will be discussed in the next section. The EPA can also necessitate the PRPs to perform the cleanup either voluntarily or involuntarily. The cleanup process falls under two categories of response actions: removal or short term and remedial or long term. Just about any action which can reduce the threat of a hazardous waste site and can be done promptly qualifies as a removal. A removal action could include providing alternate water supplies, immediate cleanup of spills from containers, or even building fences around a hazardous area. In order for the site to be eligible for removal it must meet certain criteria. The removal action must be completed within one year and must cost no more than two million dollars. Remedial actions differ in that they usually take years, even decades to complete and millions of dollars. Some examples of remedial actions could include the construction of dikes, trenches, or clay covers; excavations; and the permanent destruction or neutralization of hazardous substances. The EPA has created a system which they use to keep track of site information, it is known as the Comprehensive Environmental Response and Liability Information System (**CER-CLIS**). CERCLIS contains an official inventory of sites designated within CERCLA and encompasses the planning and tracking processes involved. The EPA gives public access to this list, allowing anyone to search for sites statewide in the CERCLIS database, the web page is <*http://www.epa.gov/superfund/sites/cursites/index.htm*>.

The first major step taken when administering remedial actions is a Remedial Investigation/Feasible Study (RI/FS). This study is the most important because it determines the size and scope of a cleanup. The Remedial Investigation includes extensive soil and ground water sampling, locating and identifying the contaminants, and defining any threats to human health and the environment. The objectives of the Feasible Study is to development remedial alternatives and evaluate in detail the potential remedies for each particular site. One of the most difficult and controversial decisions the EPA has to make is in determining the appropriate level of cleanup. Each site's relationship with the public and the environment is different. The real questions to be answered are whether the sites are to be cleaned up to pristine, predisposal conditions, or if certain levels of contaminants can remain without causing harm to humans or ecosystems. [2] The Housatonic River and the General Electric plant in Pittsfield, Massachusetts are good examples of CERCLA in action. The EPA, along with the Massachusetts Department of Environmental Protection (**DEP**), have successfully gotten GE to accept its corporate responsibility to remove and remediate soils contaminated with PCBs. General Electric has pledged in a consent decree $45 million for the cleanup and revitalization of the surrounding area.[15] Within this cleanup and revitalization process GE has agreed to, amongst others, the following actions:

1. Remediation of surface soils to levels that will allow for commercial, industrial and recreational uses.
2. Conduct flood plain remediation which will include the use of biological indicators as a mean of protecting and evaluating water quality.

3. Along with remediation of wetland sediment, a nearby brook will be rerouted and its sediments removed.
4. Extensive groundwater monitoring systems and a leachate collection system will be combined with engineering controls and long term monitoring to control the risk of exposure created by large storage areas containing contaminated material.[16]

In order for the cleanup remedy to begin, it must be subject to certain public and administrative reviews. After the completion of the RI/FS, the EPA is required to present a Record of Decision (**ROD**), which sets forth the selected remedy as well as the factors which led to its conclusion. The ROD must clearly illustrate all facts, analyses of facts, and site-specific determinations in each situation. The ROD must respond to any public comments and also be placed in an administrative record supporting the EPA's actions at a site. The EPA is required to compile a record of the entire remedial process which is subject to review and comment from both the public as well as the state. The administrative record is critical, not only in the decision making process, but also to any subsequent judicial review of the EPA's preferred remedy. Unless review of the administrative record shows the EPA's decision to be arbitrary and capricious, the decision will be upheld.

The completion of the ROD is one of last hurdles to clear before any remediation can actually begin. Its completion allows for the EPA's remedial design (**RD**) to become a detailed design permitting the construction, operation and maintenance of the project. The last stipulation is that the EPA must enter into a contract or agreement with the state which will provide future maintenance and assure payment of ten percent of the cost of remediation. The cost to implement a remedy at a CERCLA site often varies immensely. They seldomly cost less than $1 million and sometimes can exceed $50 million.[2]

The EPA's Enforcement Policy

Before the EPA can initiate an enforcement action, they must first identify the party/parties responsible for the site's contamination. Because many CERCLA sites are the result of disposal activities by hundreds of companies, when the EPA is finding who is liable, the courts have found that CERCLA imposes strict, joint, and several liability. This enables the EPA to be able to seek out companies without any requirement that their hazardous wastes have ever caused a response action or cleanup. CERCLA's standard for causation is minimal. In fact, in some cases there is arguably no causation requirement with regard to individual defendants of multiparty sites. It has typically not mattered at multiparty sited, whether a party's own waste was released as long as some waste had been discharged.[2]

The initial search for responsible parties follows a highly structured procedure. It consists of a search for the Potentially Responsible Parties (**PRPs**). This involves obtaining and organizing all available documents such as invoices and manifests associated with the sites operations. The EPA uses Section 104 (e) of CERCLA, which authorizes the EPA to attain all relevant information. This information may include the nature an quantity of materials disposed of at the site, the nature and extent of any releases at the site, and information concerning a company's ability to pay for cleanup. Penalties for failure to comply with any request can amount to $25,000 per day. A liable party under Section 107 (a) can generally be viewed as any party having some involvement with the creation, handling, or disposal of hazardous wastes at a site. The categories of liable parties include: current owners and operators, former owners and operators and parties that have, at some time, disposed of, treated, or transported hazardous substances found at a contaminated site.

The easiest to identify are current owners and operators. They are held liable with or without having been involved in waste disposal during its period of ownership and operation. With the 1986 SARA amendments, the innocent purchaser clause was created. This is viable only when the current owner or operator can prove that they had no prior knowledge of disposal. The party must show that it investigated previous ownership and property uses prior to purchase. The liability provisions concerning former owners and operators are distinctly designed to reach to prior owners at the time of disposal. This broad interpretation creates liability even if no wastes were disposed during ownership. It has been suggested that continued migration of hazardous substances can constitute discharge. Generators and arrangers make up a group of liable parties defined as: any person, who by contract, agreement, or otherwise arranged for disposal or treatment of hazardous substances which are later found to have contaminated a site. This group is often a focus of concentration for CERCLA because of the fact that most are Fortune 500 companies with the "deepest pockets," which is beneficial to the funding of cleanup processes. The courts have broadly interpreted this as being practically any situation between two entities involving handling or disposal of wastes, without necessarily intended or knowing that improper disposal would occur.[2]

Settlements with the EPA

There have been some cases involving CERCLA which have proceeded through trial but these are exceptions not the rule. More often than not settlement is the norm in CERCLA cases, and is preferred to conserve Superfund monies and allow EPA personnel to focus on other important business. PRPs also benefit by avoiding the tremendous costs of litigation and are allowed more control over selection and implementation of remedial actions as a result of settling. Settlements are ordinarily finalized in either a Consent Decree or an Administrative Order (**Consent Order**). The difference between the two forms of agreement is that a Consent Decree is filed with and signed by a federal court, while a Consent Order does not involve judicial action.

At most multiparty sites there are a large number of companies who have disposed of relatively small quantities of hazardous substances, these are so-called *de minimis* parties. *De minimis* settlements are appropriate in situations when the amount and toxicity of discharged material is low compared to others. Aside from the opportunity for an early settlement, *de minimis* parties are usually offered settlements in real finality. This means that *de minimis* parties will not be required to participate in or fund future remediation efforts.[2]

The Future of Superfund

The Environmental Protection Agency's Superfund program is the most aggressive hazardous waste cleanup program in the world. Everyday, Superfund managers are involved in critical decisions that affect public health and the environment. [7] Since the program's inception, the EPA has evaluated more than 40,000 sites, conducted more than 5000 cleanups, and restored more than 600 of the nations worst contaminated NPL sites. They use the best available science to determine risks at sites. New technologies are being developed to help achieve faster less expensive ways to cleanup sites.[17] These innovative technologies, such as bioremediation and soil vapor extraction, offer less invasive and more cost-effective alternatives. In addition, preferred technologies for sites with similar characteristics have been developed using historical patterns of remedy selection.[13]

President Clinton, in February 2000, requested $7.3 billion for the U.S. Environmental Protection Agency in the fiscal year 2001. This represents the largest budget for the agency in his eight year administration. The proposed budget earmarks $1.45 billion to continue the cleanup of the nation's worst Superfund sites. More government support, coupled with further development of initiatives such as innovative technologies, wisdom, and experience, could allow the Superfund program to make even greater strides in meeting the challenges of the future.[18]

REFERENCES

1. **Environmental Health and Safety,** *Typical contaminants found on industrial properties,* Environmental Health and Safety, Online (EHSO), Copyright 1999-2,000, *http://www.ehso.com/contaminants.htm,* updated May 6, 2000.

2. **Miller, M. L., Esq., Miller, B., and Freer, B.**, Releases or threats of releases, *Environmental Law Handbook*, Fifteenth edition. 1999, 352-380.

3. **Allan, S. M.,**. *What happened to Love Canal? http://cems.alfred.edu/student98/allansm/index.html,* "The Love Canal disaster: error in engineering or public policy", Joshua Hertz, *http://ethichs.cwru.edu/contest/canal/lovecanal.html*, 1999.

4 **Vance, B., Weinsoff, D.J., Henderson, M.A., and Elliot, J.F.**, *Toxics Program Commentary Massachusetts*, prepared by Touchstone Environmental, Inc., and published by Specialty Technical Publishers, Inc. North Vancouver, B.C. Canada V7m 1a5, 1992.

5. **U.S. Environmental Protection Agency (EPA)**, *CERCLA Overview*, *http://www.epa.gov/superfund/whatissf/cercla.htm*, Updated February 21, 2000.

6. **Contrada, F.**, *Pittsfield, sensing betrayal, confronts GE's toxic legacy*, Union News , Hampshire/Franklin County, 4/8/A12, 1999.

7. **U.S. Environmental Protection Agency (EPA)**. Citizen's Guide to Superfund. *http://www.epa.gov/reg3hwmd/super/sfguide.htm,* Updated May 20, 1999.

8. **U.S. Environmental Protection Agency (EPA)**, National Contingency Plan, *http://www.epa.gov/oilspill/ncpover.htm*, Updated March 1, 1999.

9. **The Legal Information Institute (Cornell Law School)**. *National Contingency Plan, http://www4.law.cornell.edu/uscode/unframed/ 42/9605.htm*, May 20, 1999.

10. **U.S. Environmental Protection Agency (EPA)**,   Emergency Planning and Community Right-to-Know Act. (EPCRA) *http://www.epa.gov/unix0008/toxics_poisons/epcra/epcraright.html*, Updated May 6, 2000.

11. **U.S. Environmental Protection Agency (EPA)**, *Hazardous Ranking System*, *http://www.epa.gov/superfund/programs/npl_hrs.htm*, Updated August 27, 1999.

12. **U.S. Environmental Protection Agency (EPA)**. *National Priorities List.* *http://www.epa.gov/superfund/whatissf/npl_hrs.htm*, Updated March 21, 2000.

13. **U.S. Environmental Protection Agency (EPA)**, *National Priorities List Progress, http://www.epa.gov/superfund/accomp/ei/progress.htm*,Updated December 1, 1998

14. **U.S. Environmental Protection Agency (EPA),** *National Priorities List Graphs, http://www.epa.gov/superfund/whatissf/mgmtrpt.htm,* Updated March 21, 2000

15. **U.S. Environmental Protection Agency (EPA),** *General Electric/Pittsfield, MA, http://www.epa.gov/region01/ge/index.html*, Updated February 14, 2000.

16. **U.S. Environmental Protection Agency (EPA)**, *Cleanup At General Electric*, *http://www.epa.gov/region01/ge/gerrra.html*, Updated October 22, 1999

17. **The Environmental News Network (ENN)**, Superfund Initiatives, *http://www.enncom/enn-news-archive/1999/08/081299/superfund_4635.asp*, August 12, 1999.

18. **The Environmental News Network (ENN)**, *Superfund Budget Increase, http://www.enncom/enn-news-archive/2000/02/02102000/epabudget2000-9904.asp,* February 10, 2000.

WEB LINKS FOR THIS CHAPTER

CERCLIS database
http://www.epa.gov/superfund/sites/cursites/index.htm

EPA query page
www.epa.gov/enviro/html/rcris_query_java.html

Introduction to EPCRA/SARA Title
http://www.epa.gov/opptintr/tri/intro.htm

Full text of CERCLA
http://www4.law.cornell.edu/uscode/unframed/42/ch103.html

Maps Showing Location of Superfund Sites
http://www.epa.gov/superfund/sites/npl/npl.htm

Massachusetts Contingency
http://www.magnet.state.ma.us/dep/bwsc/regs.htm

Massachusetts Oil and Hazardous Material Release Prevention and Response Act
http://www.state.ma.us/legis/laws/mgl/gl-21E-toc.htm

National Contingency Plan (NCP)
http://www.epa.gov/oerrpage/oilspill/ncpover.htm

NCP Key
http://www.epa.gov/oerrpage/oilspill/ncpkeys.htm

Summary of CERCLA
http://www.epa.gov/reg5oopa/defs/html/cercla.htm

EPCRA
Sara Title III

4

Maggie Wood

EPCRA: EMERGENCY PLANNING AND COMMUNITY RIGHT-TO-KNOW ACT (SARA Title III)

In the middle of the night on December 3, 1984, over 40 tons of methyl isocyanate (MIC) and other lethal gasses leaked from the American corporation Union Carbide's pesticide factory in Bhopal, India. According to the Bhopal Peoples Health and Documentation Clinic (**BPHDC**) 8000 people were killed in its immediate aftermath and over 500,000 people suffered from injuries.[1]

The tragic events that took place in Bhopal, India initiated the signing of the Emergency Planning and Community Right-to-Know Act (EPCRA) of 1986, also known as Title III of the Superfund Amendments and Reauthorization Act (SARA), which amended the Comprehensive Environmental Response, Compensation and Liability Act (CERCLA). The main objectives of EPCRA were to provide the public access to information concerning hazardous chemicals present in the community, and to improve emergency planning and notification.

EPCRA is made up of two components: Emergency Planning and Community Right-to-Know. The former part requires communities to develop local emergency response plans to implement in the event of a hazardous chemical release, while the latter part is aimed at increasing public awareness of chemical hazards in their community and allowing the public and local governments to obtain information about these hazards.

EPCRA is broken down into subchapters I through III. Subchapter I establishes the framework for state and local emergency planning. Subchapter II sets forth reporting requirements for the submission of information to federal, state, and local agencies concerning the type and quantity of hazardous chemicals maintained at certain facilities, and subchapter III contains miscellaneous provisions concerning trade secret protection, enforcement, and public availability of information.[2]

Note that a full text of EPCRA can be found at the following Web site [3] *<www4.law.cornell.edu/uscode/42/ch116.html>*.

Subchapter I: (Sections 301-305)

State and Local Emergency Planning and Notification [2]

The governor of each state is required to appoint a State Emergency Response Commission (SERC). SERC must designate local emergency planning districts, set up Local Emergency Planning Committees (LEPCs) within these districts, and supervise and coordinate the activities associated with each of the LEPCs. SERC is responsible for reviewing local emergency plans and must also process requests from the general public for EPCRA information.

Local Emergency Planning Committees (LEPCs)-These Emergency Planning Districts may consist of existing political subdivisions or multi-jurisdictional planning commissions. Each committee is made up of at least one member representative from each of the following categories: public elected state and local officials, police, fire, civil defense, public health professionals, environmental, hospital and transportation officials, representatives of facilities subject to Subchapter I requirements, community groups, and media representatives. The primary responsibility of each LEPC is to develop an emergency response plan and evaluate available resources for preparing for and responding to a potential chemical accident. In addition, the committee is required to give public notice of its activities, and establish procedures for handling public requests for information.

Subchapter II: (Sections 311-313)

Reporting Obligations

EPCRA created four types of reporting obligations for facilities that store or manage specified listed chemicals.

Notification of Extremely Hazardous Substances [2]

EPCRA §302 requires facilities to notify the SERC and the LEPC of the presence of any "extremely hazardous substance" (EHS) if it has the substance in excess of the specified "threshold planning quantity" (TPQ). The list of such substances is in 40 CFR Part 355, Appendices A and B. It also directs the facility to appoint an emergency response coordinator.

Amendments to the Hazardous Chemical Reporting Thresholds can be found at *<www4.law.cornell.edu/uscode/42/ch116.html>*.[4]

Notification During Releases [2]

EPCRA §304 requires facilities to notify the SERC and the LEPC in the event of a release exceeding the "reportable quantity" (RQ) of a CERCLA hazardous substance or an EPCRA extremely hazardous substance. EPCRA extremely hazardous substances and reportable quantities are listed in 40 CFR 355.

The following agencies may be notified in the event of such a release: state police, fire department, national and state emergency response agencies, state and district Title III agencies. A written follow-up report is subsequently required.

Emergency Planning [2]

EPCRA §311 and §312 require facilities to notify SERC, LEPC, and the local fire department of all hazardous chemicals for which the Occupational Health and Safety Administration requires material safety data sheets (MSDSs). All facility owners or operators are required to prepare or have available MSDSs for on-site hazardous chemicals at or in excess of established thresholds. The facility must submit either the MSDSs or a list of the substances for which MSDSs are maintained.

EPA reporting thresholds are the same for both Section 311 and 312 requirements. MSCS information and annual chemical inventories must be reported and submitted for all on-site hazardous chemicals present at any time in quantities equal to or greater than the following:

- for hazardous chemicals-10,000 pounds
- for EHS-the lesser of 500 pounds or the chemical's TPQ

MSDS Reports §311 [2]

As of October 17, 1990, businesses are required to provide either an MSDS for each on-site chemical in excess of its threshold quantity, or a list of all such chemicals to state and district Title III agencies and to the local fire department.

Facilities are allowed 3 months to prepare and submit forms, as new chemicals become subject to the MSDS requirement. If a list is submitted, hazardous chemical inventory forms (also known as Tier I and II forms) must also be submitted.

Chemical Inventories §312 [2]

Facilities subject to MSDS requirements must file annual "emergency and hazardous chemical inventory statements." Planning agencies can request either summary "Tier I" chemical inventory information, or more detailed "Tier II" information. Reports are due by March 1 of each year; however, some state and local agencies have different deadlines.

Available Information (Public Access)

1. Any member of the public can request Tier I or II information under Title III provisions.
2. Agencies must provide information concerning regulated on-site chemicals in quantities in excess of 10,000 pounds. However, before providing information about smaller chemical quantities, the agencies may require justification from the requestor.
3. Agencies are only required to ensure public access to MSDS information.

A "Tier I" form provides information about hazardous chemicals grouped by hazard category. A "Tier II" form provides information about each specific hazardous chemical. This information helps the local government respond in the event of a spill or release of the chemical. These requirements are found at 40 CFR 370, Hazardous Chemical Reporting: Community Right-to-Know.

Toxic Release Inventory [2]

EPCRA §313 requires manufacturing facilities included in SIC codes 20 through 39, which have ten or more employees, and which manufacture, process, or use specified chemicals in amounts greater than threshold quantities, to submit an annual toxic chemical release report to EPA. This report, commonly known as the Form R, covers releases and transfers of toxic chemicals to various facilities and environmental media, and allows EPA to compile the national Toxic Release Inventory **(TRI)** database. These requirements can be found at 40 CFR 372, Toxic Chemical Release Reporting: Community Right-to-Know.

Form R [2]

"Form R" reports must include the following information for the current and previous calendar years:

1. estimated quantities of both routine and accidental releases of listed "toxic chemicals;"
2. the maximum amount of listed on-site chemicals during the calendar year;
3. the amount of listed chemicals contained in wastes transferred offsite.

Report Deadlines

These reports estimating releases during the previous calendar year are due each July 1. (The present list totals 317 individual chemicals and 20 additional groups of related chemicals.)

Pollution Prevention Act [2]

In October 1990, Congress signed the Pollution Prevention Act, requiring all facilities filing annual toxic chemical release reports (Form R) under Section 313 to submit reports on toxic chemical source reduction and recycling activities for the previous calendar year. These reports (contained in Form R Part II, Section 8) must include the following information reported for the current and previous reporting years, and estimated for the next two reporting years:

- the total quantity of toxic chemicals released, including; any spilling, leaking, pumping, pouring, emitting, emptying, discharging, injecting, escaping, leaching, dumping, or disposing into the environment;
- the quantity used for energy recovery onsite and offsite;
- the quantity recycled onsite and offsite;
- the quantity treated onsite and offsite.

The report must also include the following for the current reporting year:

- the amount of toxic chemicals released as a result of any one-time events (such as an accident or remediation project);

- the ratio of the reporting year's production to the previous year's production;
- any source reduction practices used;
- an indication whether additional optional information on source reduction, recycling, and pollution control activities is included with the report.

Subchapter III: (Sections 321-330)

Trade Secrets [2]

According to Section 322, a person required to submit information under Sections 303, 311, 312, or 313 may withhold from any such submission a specific chemical's identity if he or she claims that the information constitutes a trade secret and explains the basis for this belief.

The information that he or she seeks to have withheld from the required submission must be submitted to the EPA so that the EPA may make a determination as to whether the information truly constitutes a trade.

Section 323 creates one exception to the confidentiality of trade secrets: a specific chemical identity must be disclosed where it is necessary for diagnosis or treatment of an individual by a health professional.

Public Access to Information [2]

The public has access to the following information:

- every emergency response plan
- MSDS lists
- inventory forms
- toxic chemical release forms
- follow-up emergency notices

Except for information protected as a trade secret

Each Emergency Planning Committee must designate a location where such information may be reviewed during normal working hours.

Enforcement [2]

EPCRA is administered at the federal level by the United States Environmental Protection Agency (EPA) which carries out the following duties:

- Receives Toxic Chemical Release Inventory (Form R) § 313 reports
- Oversees Title III implementation worldwide
- Establishes and revises the lists of hazardous chemicals
- Fines companies for noncompliance

Civil Actions

Citizen Suits [2]

Any person may file a civil action against a facility owner or operator for the following violations:

- failure to submit a follow-up emergency notice under § 304
- failure to submit a MSDS or list under § 311
- failure to complete and submit an inventory form under § 312
- failure to complete and submit a toxic chemical release form under § 313

Any person may bring a civil action against the Administrator of the EPA for any of these violations:

- failure to publish inventory forms under Section 312
- failure to respond to a petition to add a chemical to or delete a chemical from the § 313 list of toxic chemicals
- failure to publish a toxic chemical release form under Section 313
- failure to establish a computer database of toxic chemicals in accordance with Section 313
- failure to promulgate the trade secret regulations required by Section 322
- failure to render a decision within nine months in response to a petition by a member of the public to review the classification of information as trade secrets

State or Local Government Suits [2]

A state or local government may sue a facility owner or operator for failure to complete any of these required actions:

- notify the emergency response commission of its being covered under Subchapter A of EPCRA
- submit an MSDS or list under Section 311
- make available information requested under Section 311
- submit an inventory form under Section 312

Executive Order 12969 [2]

Mandating that each federal agency include in contract solicitations, as an eligibility criterion for competitive acquisition contracts expected to exceed $100,000, the requirement that federal contractors ensure that Toxic Chemical Release Inventory Forms (Form R) are filed by their covered facilities for the life of the contract.

Internet Resources

The most complete web site on the internet containing information on EPCRA can be found at <*www.epa.gov/swercepp/crtk.html*>[6], hosted by the Chemical Emergency Preparedness and Prevention Office (CEPP) of the EPA. This web site contains information on EPCRA reporting, legislation and regulations, documentation, chemical data and external resources. It also contains links to other useful EPCRA-related sites.

REFERENCES

1. **Corporate Watch**, *Bhopal and the WTO*, Nov. 24, 1999.
   <http://www.igc.org/trac/bhopal/global.htmll>

2. **Von Oppenfeld, R. R.,** *"Emergency Planning and Community Right-to-Know Act,"* Environmental Law Handbook, 15th Ed., Sullivan et al., Government Institutes, 1999, 617-643.

3. **Legal Information Institute**, U.S. Code, Title 42, Chapter 16, Emergency Planning and Community Right-To-Know
   . <www4.law.cornell.edu/uscode/42/ch116.html>

4. **EPA**, Federal Register (Volume 64, Number 28) Feb 11, 1999.
   <http://www.epa.gov/fedrgstr/EPA-TRI/1998/June/Day-08/tri14490.htm>

5. **EPA**, Federal Register (Volume 63, Number 109) June 8, 1998.
   <http://www.epa.gov/fedrgstr/EPA-GENERAL/1999/FEBRUARY/Day-11/g3255.htm>

6. **EPA**, Chemical Emergency Preparedness and Prevention Office, Preparedness-Emergency Planning and Community Right-To-Know, 2000.
   <www.epa.gov/swercepp/crtk.html>

WEB LINKS FOR THIS CHAPTER

EPCRA Summary
http://www.epa.gov/region5/defs/html/epcra.htm

Full Text of EPCRA
http://www4.law.cornell.edu/uscode/42/ch116.html

Amendments to Hazardous Chemical Reporting Thresholds
http://www.epa.gov/fedrgstr/EPA-TRI/1998/June/Day-08/tri14490.htm

Consolidated List of Lists
http://www.epa.gov/ceppo/pubs/title3.pdf

SARA Title III Rules Page
http://www.epa.gov/swercepp/rules/epcra.html

EPCRA Homepage
http://www.epa.gov/region04/air/epcra/epcrtext.htm

EPA Office of Solid Waste and Emergency Response
http://www.epa.gov/swerrims

TRANSPORTATION OF HAZARDOUS MATERIALS

5

Michael Wood

SUMMARY OF REGULATIONS

The transportation of hazardous materials can pose an immediate threat to the health and safety of people and the environment. Regulations and requirements controlling the transportation of hazardous materials are necessary to prevent an unwanted release to the environment, or an exposure to any persons who might come in contact with a hazardous material. The heart of the regulations pertaining to the transportation of hazardous materials is the Hazardous Materials Transportation Act (HMTA) of 1974.

The Hazardous Materials Transportation Act increased the Department of Transportation's (DOT) authority on the safe transport of hazardous materials. This act pertains to transportation by rail, highway, air, or sea. These regulations were revised in 1990 to reduce the differences of transportation requirements on the global level. The DOT achieved this by creating a numbering system under HM-181. These latest revisions were designed to stop variations in labeling between states and/or countries.

THE DEPARTMENT OF TRANSPORTATION AND THE HAZARDOUS MATERIALS TRANSPORTATION ACT

The Hazardous Materials Transportation Act of 1974 regulates the tasks of people that transport hazardous materials. These regulations control several aspects of transportation. The manufacture of transport containers, along with the labeling of containers, is heavily regulated. Placards and markings must be clearly illustrated on the container to be shipped and the vehicle that is shipping it. The personnel that are handling the material must undergo specialized training. For further information on training courses and education for the safe shipping of hazardous materials visit the Hazardous Materials Advisory Council Home Page. The web site is *http://www.hmac.org/*. Hazardous materials must be registered before they are shipped. Certain hazardous materials must have special transport routes designated for them. Transport routes can also be restricted to certain materials. In the case of a release of hazardous materials the spill must be reported. The transporter also has financial responsibility for the materials that are transported in the instance of a release.[1]

The Department of Transportation (DOT) oversees and also regulates the HMTA and classifies hazardous materials according to their greatest potential hazard. The DOT defines a hazardous material as anything that presents a threat to health, safety, and the environment when transported. The secretary of transportation decides what is classified as a hazardous material. Hazardous products are also regulated by the EPA under 49 CFR 172.101.[1] The DOT web site can be found at *http:// www.dot.gov/rules.htm*.

An update to the original HMTA of 1974 was proposed on February 16, 1999. The 1999 Hazmat Transportation Safety Reauthorization Act was brought to Congress by the Secretary of Transportation. This reauthorization act is very similar to the one proposed in 1997. The reauthorization act of 1999 is designed to increase the authority of the Department of Transportation involving the enforcement of transportation laws. The reauthorization act will expand the effectiveness and usefulness of this program. The reauthorization act of 1999, the 1997 proposal, and a section-by-section analysis of this proposal can be found on the DOT Office of Hazardous Materials and Safety web site at *http://hazmat.dot.gov/99reauthact.htm*.

The DOT has designed its own hazard classification system for the easy identification of hazardous materials. This system uses a series of numbered and color coded placards, labels, and markings. These placards identify a hazardous material's most threatening hazard, like flammability or radioactivity for example. This can help an emergency response team identify what type of material has been released. The DOT's hazard classification system separates hazardous materials into nine hazard classes, with some subclasses. This system is set up with the first class being the most dangerous class (explosives) all the way to a miscellaneous section, which could cover anything that does not fit hazard classes one through eight. These placards have to be displayed on the transport vehicle in designated areas, and they must be a special size. The diamond shaped placards must have a four-digit ID number on it or in an orange rectangle below the placard. The four-digit number is the UN number, which denotes the specific chemical in transport. These numbers can be obtained for cross-reference in the North American Emergency Response Guidebook also known as the DOT pocket guide. The placard must be displayed on the ends and sides of the cargo tank, vehicle, or rail car. [3]

Class 1-Explosives:

Explosives are any materials that can have a rapid pressure build up and release heat and or gases. These materials can also react with themselves and result in an explosion. Explosives are most likely unstable and must be shipped with extreme care. An example of an explosive would be dynamite.

Class 2-Gases:

Gases are materials that can be displaced or expand in areas very rapidly and could

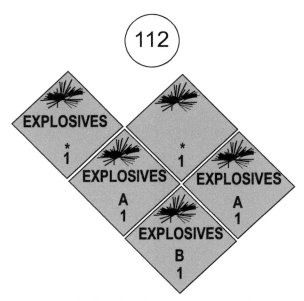

Figure 5-1. DOT Hazard Classification System. Placards for Class 1- Explosives.
Source: adapted from CANUTEC Home Page. [6]

cause serious injury or death. Most of these materials are also called compressed gases, because they are shipped under extreme pressure. In a fire, as the temperature rises the pressure in the tank rises and could rupture the tank. This condition is also called BLEVE, or boiling liquid expanding vapor condition. These compressed gases are divided into 3 class divisions.

Division 2.1 Flammable Gases: These gases will boil at 20^{0}C and 101 kPa of pressure. They are ignitable in 13% or less oxygen. These compounds have a 12% flammability range despite the Lower Explosive Level (LEL).

Division 2.2 Non-flammable Gases: Non-flammable gases are shipped under extreme pressure because they are compressed into somewhat small heavy-duty tanks. The material has an absolute pressure of 280kPa at 20^{0}C or higher. They are not put in division 2.1 or 2.3.

Division 2.3 Poisonous or Toxic Gases: These are gases that are extremely toxic to life and health. They have a boiling point of 20^{0}C or less, and a pressure of 101.3 kPa. The toxicology of these gases gives an LD50 of about 5000ml/m^3.

Class 3-Flammable Liquid:
  Flammable liquids are materials that have a flashpoint of no more than 141^{0}F (60.5^{0}C). This includes liquids of no more than 100F (37.8^{0}C) and are heated and transported at or above the flashpoint. The flashpoint is defined as the temperature in which

Figure 5-2. DOT Hazard Classification System. Placards for Class 2-Gases.
Source: adapted from U.S. DOT [3,4]

a material produces a vapor that will ignite in the ambient air. An example of this would be gasoline.

Class-4 Flammable Solids:
   Flammable solids are separated into 3 specific divisions for easier identification. Flammable solids can be materials like smokeless powder for small arms, or sodium metal.

Division 4.1 Flammable Solids: Flammable solids are materials like explosives that have been wetted to stabilize them. These materials can also be self-reactive or highly combustible.

Division 4.2 Spontaneously Combustible Materials: These materials are considered pyrophoric, or they ignite when exposed to the ambient air.

Division 4.3 Materials Dangerous When Wet: These materials are hazardous when wet because they undergo ignition when exposed to water. This type of material is important to know about in an emergency response situation, especially when there is a fire. Spraying this material with a fire hose could be disastrous.

Figure 5-3. DOT Hazard Classification System. Placards for Class 3 Flammable Liquids. Source:aadapted from U.S. DOT [3,4]

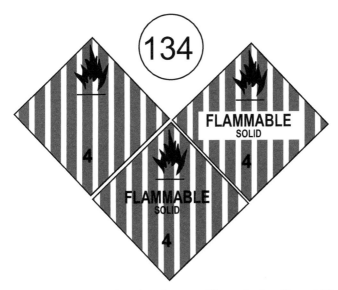

Figure 5-4. DOT Hazard Classification System. Placards for Class 4 Flammable Solids. Source: adapted from U.S. DOT [3,4]

Class 5-Oxygen Containing Materials

Division 5.1 Oxidizers: These materials use oxygen to enhance combustion that is created by other materials or create a fire themselves.

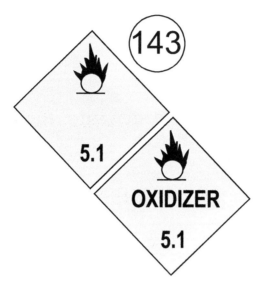

Figure 5-5. DOT Hazard Classification System. Placards for Class 5- Oxygen Containing Materials. Source: adapted from U.S. DOT [3,4]

Figure 5-6. DOT Hazard Classification System. Placards for Class 6- Poisons. Source: adapted from U.S. DOT [3,4]

Division 5.2 Organic Peroxides: Organic peroxides are peroxides that have oxygen in their molecular structure, and are a derivative of hydrogen peroxides.

Class 6-Poisons
Division 6.1 Poisonous Materials: They are hazardous materials that are toxic to humans when they are transported. These compounds or chemicals can exhibit acute oral toxicity, dermal toxicity, and/or acute inhalation toxicity in humans.

Division 6.2 Poisonous Materials: Infectious Substances: These materials are usually medical diagnostic specimens, biological products, and medical waste.

Class 7-Radioactive
   This type of material is able to emit invisible and harmful radiation.[3] Radioactive materials are any substances that have a radioactivity more than 0.002 mCi/g. This waste is very carefully and stringently regulated and specific information can be found in 49 CFR 173.401-173.478, and also 10 CFR 71.[1]

Class 8-Corrosive Materials
   This type of material can cause destruction and irreversible damage to the tissue of humans when they are exposed to the hazardous substance. These materials are also capable of corroding steel and aluminum.[1]

Figure 5-7. DOT Hazard Classification System. Placards for Class 7-Radioactive.
Source: adapted from U.S. DOT [3,4]

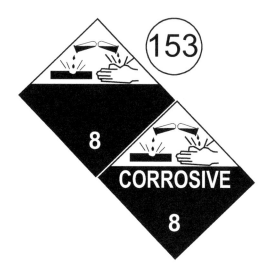

Figure 5-8. DOT Hazard Classification System. Placards forClass 8-Corrosives.
Source: adapted from U.S. DOT [3,4]

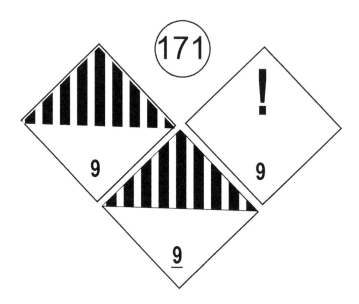

**Figure 5-9. DOT Hazard Classification System. Placards forClass 9
Miscellaneous.** Source: adapted from U.S. DOT [3,4]

Class 9-Miscellaneous

This is the final class of hazardous materials and it contains items that can create anesthetic or noxious properties that can cause great discomfort and annoyance. This is also the category where all other regulated hazardous materials are placed.[1]

Exceptions and Specific Regulations of the Hazard Classification System

The DOT Hazard Classification System undergoes changes to the individual classes as regulations change. New divisions are added and changed from the classification system. There are different divisions in other countries like Canada. There are 3 divisions observed in Canada in the Miscellaneous (9) class. They are as follows:

1. Miscellaneous Dangerous Goods
2. Environmentally Hazardous Subsrances
3. Dangerous Wastes

There is also a Division 2.4 in Canada under the Gases Division (2), which is listed as Corrosive Gases. The classification system also breaks down explosives into 6 divisions. They are described as:

1. Explosives with a mass explosion hazard
2. Explosives with a projection hazard
3. Explosives with predominately a fire hazard
4. Explosives with no significant blast hazard
5. Very insensitive explosives; blasting agents
6. Extremely insensitive detonating articles

The Hazard Classification System only requires the hazard to be written on the placard in the U.S. Class 7 and Oxygen placards do not require the primary hazard to be written on the placard. These exceptions and specific regulations can be found in a current DOT Guidebook. This information can also be obtained at the EHSO-Environmental Health and Safety Online web site: *<http://www.ehso.com/DOTHow2Comply.htm>*.

Shipping Hazardous Waste: Exceptions to the Rule

There are exceptions to every regulation and the same goes for the shipment of hazardous materials. The exceptions that are described are just a few of many specific regulations under CFR title 49. These exceptions pertain to the traditional DOT shipping, labeling, and placarding requirements. The materials under this specific regulation are hazardous materials that fall under DOT Hazard Classifications 3, Divisions 4.1, 4.2, 4.3, 5.1, 6.1, and Classes 8 and 9. If the material is poisonous when inhaled it is not to be shipped under the requirements of this section. Rail, motor vehicle or cargo aircraft under this section may ship these materials.[2]

Liquids that pertain to this section are to be shipped using and abiding to these requirements. Liquids must be packaged in a glass, plastic or metal lined container not to exceed 1.2 liters. The container must be tightly sealed. The net quantity of the liquid at 20 degrees Celsius cannot exceed 1 liter. The inner packaging of the container must be hermetically sealed in a bag, and there must also be an absorbent material that will not react with the product that is to be shipped. This absorbent must absorb the product in the event of a release. The absorbent material has to fit tightly in the outer container. The metal container must be closed tightly and hermetically sealed for Divisions 4.1 and 4.2. The metal can is to be put into a fiberboard box that has hermetically sealed packing around the product. The inner packing must be put in secure outer packing according to Section 173.201(under 49 CFR). Only four intermediate packages are allowed in one outer packing. [2]

Exceptions designed for packaging solids involve different methods for safe transport. The materials must be placed in tight plastic, metal, or glass containers. The net weight of the container cannot exceed 2.85kg. There must be an impervious hermetically sealed barrier bag around the product. The barrier bag must be placed with its contents in a fiberboard box that is hermetically sealed around the box. The inner packaging must be placed in secure outer packaging that conforms to Section 173.211 (under 49 CFR). The maximum amount of intermediate packages that can be put into one outer packaging is four. The outer packaging must be labeled with the proper shipping name that says, " This package conforms to 49 CFR 173.13." [2]

There are several specific regulations to certain materials and how they are to be shipped. These special regulations are found in 49 CFR 173 "Part 173—Shippers—General Requirements for Shipments and Packagings", and can be found at the web site: <*http://www.access.gpo.gov/nara/cfr/waisidx_98/49cfr173_98.html*>. [2]

Packaging Hazardous Waste and Materials

The DOT began packaging requirements to comply with the United Nations (UN) regulations in 1990. Hazardous Materials are packaged using a general system that conforms to the hazard characteristics of a given substance. This creates an alternative to material specific packaging that is far too complicated. Guidelines utilized in packaging include compatibility, leak and spill protection, vibration resistance, and temperature changes. The UN packaging system breaks down hazardous materials into three groups. These are groups designated by Roman numerals I, II, and III. Packing group I is the most hazardous, while packing group III is the least hazardous. The DOT created new regulations that state packaging must comply with 1994 regulations. [1]

General requirements for packing and shipping involve strict regulations. The methods for packing, shipping and storing hazardous materials require inspection by an authorized representative. The regulations set forth for specific materials can be applied to all forms of transportation unless otherwise stated. Salvage drums are required for any leaking packages. The leaking product must be placed in a removable head salvage drum that meets shipping requirements.

The salvage drum must meet UN 1A2 specifications set for packing group III for liquids and solids. The drum should include the proper cushioning and absorption material. Every drum should include the proper shipping name and names of the hazardous materials in the packaging. The drums are to be labeled for the material to be packaged and the shipper must prepare shipping papers for the product. Overpack requirements don't apply to drums in Section 173.25. A salvage packaging marked T should be supplied in accordance with UN recommendations. Further regulations on packaging can be found in 49 CFR 173 "Shippers—General Requirements for Shipments and Packaging" at the web site: *<http://www.access.gpo/gov/nara/cfr/waisidx_98/49cfr173_98.html>*.[2]

Shipping Papers

Hazardous materials that are transported must have the proper shipping papers with them. There are important requirements for the papers that accompany the shipment. Shipping papers must include:

1. The shipper's name
2. The name of the materials to be shipped
3. The hazard class.
4. The four-digit ID numbers under the UN or the NA system
5. The material packing group
6. The amount of material
7. The proper 24-hour emergency response phone number

The shipper and the manufacturer hold the sole responsibility on the accuracy of the shipping papers. A hazardous waste manifest must also accompany the shipment when in transport. [1]

Markings, Labels, and Placards. The containers in which hazardous waste is shipped must be identified clearly and labeled with the correct shippers name, ID number, and the cosigner's name and address. The diamond shaped placards must be put on all packages that identify the hazard class of the material.

The transport vehicles must display color keyed and numbered placards. The numbers show the hazard class of the material. The placards must be of an approved size and they are to be displayed on the ends and sides of the vehicle. Placard requirements are constantly changing to include different hazard class numbers. The placards and labels have to remain on the transport vehicles until the hazardous materials are removed. [1]

Mailing Hazardous Materials [7]

Hazardous waste or materials that are mailed fall under the category of Other Regulated Materials (**ORMs**). This system was created by the DOT and is listed under 49 CFR. Many people and agencies are familiar with this system and have adopted the requirements for mailing hazardous materials. Specific Regulations on ORMs and mailing hazardous waste can be found in 49 CFR 173, and at the web site *<http://www.access.gpo/gov/nara/cfr/waisidx_98/49cfr173_98.html>* .

Hazmat Technicians at Emergency Response

Emergency Response Information. Emergency response information is provided to reduce the unknown dangers of a harmful spill. This helps to create a quick, effective and safe response to a spill. The following information must be included for emergency response:

1. The description and technical name of the material
2. The immediate health hazards
3. The fire and explosion information
4. The precautions to be taken immediately in an emergency response
5. Any immediate fire control methods
6. The spill control methods that don't involve fire and explosion
7. First aid measures

The emergency response information must be stored in the transport vehicle away from the hazardous materials. A good place for this information is the cab of the transport vehicle. The transport vehicle can also carry the DOT guidebook that has the above information for thousands of materials. The transporter must carry a 24 hour emergency response phone number. [1]

DOT Guidebook[4]

The *DOT Guidebook*, also called the *North American Emergency Response Guidebook*, was created to make the identification of hazardous materials quick and easy in an emergency response. Firefighters, police, and any other emergency response teams that are the first to arrive on the scene of a transportation incident involving hazardous materials would utilize this book.[4]

This book is broken down into color-coded sections that acts as a cross-reference guide for the identification of hazardous or dangerous materials. There is also a section on how to deal with certain types of chemicals in an emergency response incident. The sections are split up into colors.

Yellow Section: The yellow section is broken down in numerical order by a UN chemical identification number. This is a very useful section when the only information that an emergency responder has is the UN number that usually appears on the placard or beside the placard in an orange rectangle. Beside the ID number is a guide number that will refer you to another section in the book. The name of the specific hazardous material is written beside the guide number. [3]

Blue Section: This section puts chemicals in alphabetical order and has the guide number and ID numbers written beside the chemical name. This would be used if the specific chemical were already known upon an emergency response incident [3]

Orange Section: The orange section of the guidebook is the section that has all of the guide numbers listed in numerical order. These numbers are referred to from the previous sections when the chemical is identified. When a guide number is identified the reader goes to the specific section. There is a section that has information for potential hazards, specifically fire and explosion data, and health data if someone has come in contact with the chemical. There is also a section on public safety. This can tell the responder how far away to evacuate the public. This part of the orange section has information on the protective clothing that should be worn when dealing with a certain chemicals. Another part of the orange section has information on how to deal with fires, spills, leaks, and what type of first aid should be administered to individuals who have been exposed to the hazardous material.[3]

Green Section: This part of the guidebook contains specific data on initial isolation and protective action distances that need to be utilized in an emergency response. This is ordered by the ID number of specific chemicals, and contains data for both small and large spills. [3]

Copies of the emergency response guidebook can be obtained from several places. There is a copy of the 2000 Emergency Response Guidebook on the web at the DOT Office of Hazardous Materials and Safety web site: <*http://hazmat.dot.gov*>.

REFERENCES

1. **Vance, B., Weinsoff, D.J., Henderson, M.A., and Elliot, J.F.**, *Toxics Program Commentary: Massachusetts*; prepared by Touchstone Environmental, Inc., Specialty Technical Publishers, Inc., North Vancouver, B.C. Canada V7M 1A5. 1992.

2. **49 CFR 173 WAIS Index,** *"General Requirements for Shipments and Packagings"* http://www.access.gpo/gov/nara/cfr/waisidx_98/49cfr173_98.html

3. **US Department of Transportation**, *1996 North American Emergency Response Guidebook*, Office of Hazardous Materials Initiatives and Training (DHM-50) Washington, D.C. 1996.

4. **DOT Office of Hazardous Materials and Safety**, *"2000 Emergency Response Guidebook"* http://hazmat.dot.gov, 2000.

5. **EHSO-Environmental Health & Safety Online** *"RSS Guide"* http://www.ehso.com/RSSGuide.htm, 1999.

6. **CANUTEC Home Page** *"NAERG96 On-Line"* http://www.tc.gc.ca/canutec/en/guide/menug_e.htm, 2000.

7. **Hazardous Materials Advisory Council Home Page,** *"Welcome to the Hazardous Materials Advisory Council"* http://www.hmac.org/, 2000.

8. **Massachusetts Department of Fire Services,** *"Hazardous Materials Response"* http://www.state.ma.us/dfs/hazmat/hazhome.htm, 2000.

WEB LINKS FOR THIS CHAPTER

DOT Office of Hazardous Materials Safety
http://hazmat.dot.gov

DOT Rules and Regulations
http://www.dot.gov/rules.htm

General Information and Regulations
http://www.access.gpo.gov/nara/cfr/waisidx_98/49cfr171_98.html

Hazardous Materials Transportation Home Page
http://www.et.anl.gov/thm/index.html

Informational Publications and Books (DOT Guidebook)
http://www.ehso.com/RSSGuide.htm
http://www.tc.gc.ca/canutec/en/menu.htm

Legislation, Regulation, Certification, and Policy Links
http://www.dot.gov/legandreg.htm

Requirements for Shipments and Packaging
http://www.access.gpo.gov/nara/cfr/waisidx_98/49cfr173_98.html

Shipping and Packaging Information
http://www.access.gpo/gov/nara/cfr/waisidx_98/49cfr173_98.html

Transportation of Hazardous Materials and Environmental Health and Safety
http://www.ehso.com

Transportation of Hazardous Materials Free Publications and Guidance
http://www.ehso.com/dotguidance.htm

Transportation Environmental Resource Center
http://www.transource.org

POLLUTION PREVENTION and WASTE MANAGEMENT

6

Rachael Weiskind

POLLUTION PREVENTION. WHAT IS IT? [1,2]

Pollution prevention, also referred to as "source reduction" means the deliberate decrease in the amount of any hazardous substance, contaminant, or pollutant that enters the environment prior to recycling, treatment, or disposal. As such, pollution prevention strategies have as their primary goal a series of employable tactics that serve to reduce or eliminate entirely the generation of pollutants and waste from the source of generation. The role that source reduction plays in the 21st century is one of ecological consequence. In addition to protecting natural and finite resources such as water and oil, pollution prevention tactics also improve the manner in which virgin resources are used. Because pollution prevention methodologies stress resource and material conservation, raw material losses and the need for expensive end-of-the-pipe technologies are greatly reduced. Source reduction therefore appears to be one of the most cost-effective strategies in fighting the war against pollution.

Industry cooperation with initiatives designed to reduce the amount of waste produced often include a variety of methodologies. Some examples of source reduction in the industry sector involve material modification or substitution, product reformation and/or on-site recycling initiatives. Additional pollution prevention measures may also include pollution reduction via improvements in the efficiency with which raw materials, energy, water, or other resources are utilized as well as the derivation of policies and regulations designed to conserve natural resources. Likewise, tactics geared to promote the safe and legal disposal of waste that cannot be prevented, reduced, or recycled are often also implemented.

A History of Pollution Prevention [1, 2]

Less than 20 years ago, the only pollution prevention effort at the Environmental Protection Agency, EPA, was a proposal that called for waste minimization activity. While a department known as the Office of Pollution Prevention and Toxics was established in 1977, the primary objective of workers in this office was to administer the statutes and guidelines of the Toxic Substance and Control Act. As such, source reduction efforts in the early 1980's were identifiable only in a limited number of industries where policies were implemented at the leisure or concern of company management.

In contrast to the former part of the decade, the latter half of the 1980s was an era that proved to be crucial in laying the foundation for future source reduction legislation. It was during this time period that many lawmakers began expressing an interest in promoting waste reduction methodologies. In response to this newly found bureaucratic awareness, the Office of Pollution Prevention and Toxics began to develop and compile information on source reduction technologies and other additional and accessory pollution prevention efforts. This increased awareness by both legislators and environmental officials alike proved to be a crucial element in the establishment and subsequent implementation of a federal mandate for source reduction in the twentieth century.

The Pollution Prevention Act of 1990 [2,3]

Passed and adopted into law in the last decade of the century, the Pollution Prevention Act of 1990, was based largely in part upon a federal investigation into the costs and benefits of source reduction. An inquiry made by Congress revealed the following facts:

1. The U.S. produces millions of tons of pollution each year and spends tens of billions of dollars in attempts to rectify the damage.
2. There are significant opportunities for industry to reduce or prevent pollution at the time of production via cost-effective alterations in production, operation, and raw material usage. Such changes would offer industries a substantial savings in raw material utilization, pollution production, and liability costs, while simultaneously working to protect the health and safety of company workers and the environment.
3. The opportunities for source reduction are often not realized because of other existing regulations, and the industrial resources that they require for compliance, focus upon treatment and disposal rather than source reduction; existing regulations do not emphasize multi-media management of pollution; and businesses need information and technical assistance to overcome institutional barriers to the adoption of source reduction practices.
4. Source reduction is fundamentally different and more desirable than waste management and pollution control.

As a direct result of these aforementioned findings, Congress appointed the EPA as the sole authority responsible for developing programs designed to help remedy the problem of waste generation. Responsibilities of this organization thus included:

1. Establish standard methods of measurement for source reduction
2. Ensure that the EPA considers the effects of its existing and proposed programs on source reduction efforts and reviews regulations of the agency prior to and subsequent to their proposal to determine the effects of source reduction
3. Coordinate source reduction efforts within each agency office and coordinate with appropriate offices to promote source reduction practices in federal agencies and generic research and development on techniques and processes which have broad applicability.
4. Develop improved methods of coordinating, streamlining and assuring public access to data collected under Federal environmental statutes.
5. Facilitate the adoption of source reduction techniques by businesses
6. Identify measurable goals which reflect the policy of the Pollution Prevention Act of 1990, the tasks necessary to achieve these specific goals, dates upon which principle tasks are to be accomplished, required resources, organizational responsibilities and the means by which progress in meeting the goals will be measured
7. Establish an advisory panel of technical experts comprised of representatives from the industry, the states, and public interest groups, to advise the Administration on ways to improve collection and dissemination of data
8. Establish a training program on source reduction opportunities (workshops etc.)
9. Identify and make recommendations to Congress to eliminate barriers to source reduction including the use of incentives and disincentives.
10. Identify opportunities to use Federal procurement to encourage source reduction
11. Develop, test and disseminate model source reduction auditing procedures designed to highlight source reduction opportunities
12. Establish an annual award/incentive program to recognize outstanding source reduction programs

The Pollution Prevention Act of 1990 established a new environmental policy for the U.S. This piece of legislation consists of five integral parts including a summarized national policy, an EPA pollution prevention strategy, grant programs, source reduction clearinghouse, and source reduction data collection. While these elements serve as the collective foundation upon which Federal environmental policy was constructed, it is nevertheless important to examine each individual component as they are defined in the Pollution Prevention Act of 1990.

National Policy [3]

Pollution should be prevented or reduced at the source whenever feasible. Pollution that cannot be prevented should be recycled in an environmentally safe manner. Pollution that cannot be recycled should be treated in an environmentally safe manner. Disposal or other release into the environment should be employed only as a last resort and should be conducted in an environmentally safe manner.

EPA Pollution Prevention Strategy [3]

Procedures are to be undertaken by the EPA to promote source reduction. Some features include a review of agency programs to determine impact on source reduction, the coordination of source reduction activities, and the identification of barriers and incentives to source reduction.

Grant Programs [3]

Grants provide funds for state programs so that they may provide technical assistance and training to businesses. Programs may target information to businesses who lack information that is important to establishing an effective program in source reduction.

Source Reduction Clearinghouse [3]

This serves as the center for source reduction technology transfer. It is used to collect and compile information from states funded by pollution prevention grants.

Source Reduction Data Collection [3]

This requires those already submitting annual toxic chemical release reports to include source reduction and recycling progress in their reports.

Grant Programs Established under the Pollution Prevention Act of 1990 [2,3]

One of the EPA's responsibilities, as stipulated by the Pollution Prevention Act of 1990, is to assist with the funding of state based source reduction programs. As such, the EPA is required to match state funds in an effort to promote pollution prevention efforts by local industries. The Pollution Prevention Incentives for States (**PPIS**) grant program is one such program that supports these federally funded pollution prevention tactics and assists with the development of state programs. This initiative was developed by the EPA to give states flexibility in addressing local needs. In short, because states have a more direct link with local businesses, a better assessment of industry wants and needs can be made, thus making federally subsidized pollution prevention efforts more comprehensive and applicable.

The PPIS program, formerly known as the Source Reduction and Recycling Technical Assistance Program, was initiated in 1989, but renamed in 1990 to emphasize the different manners in which source reduction practices could be implemented. PPIS has supported over 100 projects and provided in excess of $20 million dollars to eligible recipients. While the primary beneficiaries of the PPIS program are state agencies, territories and possessions of the United States including federally recognized Indian tribes are also eligible for these same federal funds. Local governments (i.e., county boards), privately funded universities, private non-profit groups and individuals are, however, not eligible for these federal subsides.

There are currently 49 states with pollution prevention programs and more than half of all states have additionally enacted source reduction legislation. PPIS continues to support activities within the states to promote further pollution prevention strategies. The most frequently cited activities include technical assistance, technical training, education and outreach, regulatory integration, demonstration projects, legislation and infrastructure, and awards and recognition.

Technical Assistance [2]

Technical assistance programs are provided by PPIS help industries identify pollution prevention opportunities. Many of these programs offer free, confidential, non-regulatory on-site pollution and waste assessments. In addition, assistance can often also be obtained via free hotlines or referrals to industry-specific publications. This type of assistance can be sought after by both small and large businesses and has proven, in many instances, to assist companies in saving money, increasing efficiency, reducing the need for new, costly disposable facilities, and helping to promote a good public image.

Once a company receives this type of support, a report detailing potential areas of rectification relative to source reduction is provided. In an attempt to further assist businesses in accessing this type of government service, some PPIS grant recipients have examined barriers that may impede businesses from seeking technical support. The Louisiana Department of Environmental Quality has, for instance, developed a survey to explore and identify barriers to action. Once issues central to this lack of participation are recognized, program objectives and directives will be adjusted to reflect concerns. The ultimate goal of this type of project thus involves a type of tailoring specific to the wants and needs of local businesses and communities.

The Benefits of Technical Assistance [2]

Since 1989 the Tennessee Waste Reduction Assistance program (WRAP) has provided service to over 200 sites. A compilation of what this type of assistance actually did for the companies revealed the following:

1. An average savings of $41,500 by each company a year.
2. A 1.3 million pound reduction of hazardous waste
3. An 8.8 million pound reduction of solid waste
4. A 91,000 gallon per day reduction of wastewater
5. An 87,000 gallon per day reduction of fresh water consumption
6. A 450,000 pound reduction of air emissions

Technical Training [2]

Because the overall goal of pollution prevention is to decrease the amount of waste generated at the source, the EPA encourages the sharing of information on source reduction techniques. PPIS grants are therefore designed to assist in this process by funding state programs that provide technical training to industry, government and students. As such, many state programs train businesses how to implement source reduction methodologies within their own walls. The aforementioned Tennessee Waste Reduction Assistance Program (WRAP) has developed and delivered several presentations on waste reduction strategies. Current figures estimate that the program has trained in excess of 12,000 people.

PPIS initiatives can also be employed to train state and local environmental officials. Simply stated, government subsides can also be used to educate further public persons on matters central to source reduction. Rhode Island, for instance, is educating leaders at its largest publicly owned water treatment facility. This particular project assists with the training of officials on issues central to pollution prevention tactics during compliance and independent waste audits.

Other crucial elements in this type of technical assistance are prompt and continual retraining initiatives. As such, PPIS matches funds that contribute to the reeducation of employers, employees, and other designated parties. Money is also often granted to established pollution prevention programs that assist with the training of novice sites. The Alabama program has, for example, been instrumental in the formation and development of source reduction programs in Vermont, New Hampshire, Iowa, Mississippi and South Carolina.

Similar to the training and retraining programs found within the business sector of this country are the programs that also serve to aid in pollution prevention efforts within schools. In many instances private businesses with already established programs will help with the development of source reduction measures at local colleges and universities. These types of programs often provide students with comprehensive training in identifying areas of rectification. One type of learning tool often employed by some of these businesses is a sort of mock inspection performed by the students and then graded by the leaders of the training initiative.

Outreach and Education

The Michigan Approach [2]

The Michigan board of education developed a guidebook and video to assist school policy makers, facility managers, teachers, and students in reducing the amount of waste and pollution put out by the school. The guidebook offers suggestions for source reduction including information of how to institute pollution prevention projects into the classroom. The guide places emphasis on the role of the students, as student involvement is key in the success of these proposals. Examples of student involvement include separating wastes in the classroom and cafeteria, as well as handling hazardous wastes carefully.

PPIS supports the development of curricula, educational videos, university courses, and student intern programs. Furthermore, PPIS also, in attempts to encourage further public adoption of pollution prevention measures, supports programs that target particular consumers and businesses. Recipients of this type of subsidy have developed newsletters, pamphlets, and other pollution prevention awareness materials. Likewise, many of these beneficiaries also sponsor workshops and conferences that revolve themes around pollution prevention.

The American Samoa Environmental Protection Agency [2]

The American Samoa Environmental Protection Agency (ASEPA) utilized PPIS funds to launch a major public education initiative in hopes of assisting residents with their waste. Past data on the island had indicated that most of the island's waste was biodegradable and thus disposed of on the land, in water, or through incineration. Materials that were, however, non-biodegradable (i.e., oils, paints and alike) were creating huge waste management issues for the area. In an effort to therefore rectify the situation, ASEPA developed a series of television programs and provided communities with information on proper waste management techniques such as recycling and source reduction.

Regulation Integration [2]

Because pollution knows no boundaries, it is important to consider the ramifications of our actions on the environment. PPIS thus encourages government agencies to integrate the source reduction effort into all areas of state environmental policy.
In an effort to deter and thereby reduce the transfer of pollution, the state of Massachusetts has developed a cross media inspection program that incorporates pollution reduction requirements into enforcement procedures. In a pilot study, state agencies trained teams to inspect facilities in air, water, and hazardous waste compliance. Pollution prevention technical assistance was provided simultaneously. Later analysis revealed positive results and the program was instituted into several other regional offices.

PPIS beneficiaries have also often implemented projects in hopes of increasing the coordination efforts of several different regulatory agencies. Some local governing bodies in the state of California have, for instance, joined to forces to illustrate that regional planning can also be an effective and more importantly efficient source reduction tool. The Washington Department of Ecology has too established means for increasing and improving communication amongst area offices.

Demonstration Projects [2]

The EPA encourages states to implement demonstration projects that test and support new source reduction initiatives. The underlying intention of these projects is to afford states the ability to learn how these innovative programs function before businesses or the federal government has the opportunity to intervene. As such, PPIS fosters demonstration projects in a multitude of areas including alternative waste generation tactics as well as community waste reduction and recycling programs. An example of such methodology has been undertaken in the state of Nevada, where researchers there have investigated alternatives in mining that yield precious metals.

As mentioned in the preceding paragraphs, the PPIS funds grants that address issues central to the transfer of pollution across different media. Some of these efforts have been large scale and occurred through partnerships such as the one between the Missouri Department of Natural Resources and the Tennessee Valley Authority. These two agencies have teamed up in an effort to assist agrichemical dealerships in reducing the amounts of pollution they produce. The project serves to identify areas in which change could be easily implemented.

Awards and Recognition [2]

In an effort to award those businesses that have initiated, modified, or changed their day-to-day practices in the name of pollution prevention, many states have assembled award programs to honor their participants. A program such as this began in New Jersey in 1988 and in 1991, 37 applications for awards and recognition were received. Canadian airlines has also made large strides in the effort for source reduction. Initially motivated by the need to improve employee morale, this airline implemented changes that revolved around improving and "cleaning up" the environment. It was only shortly thereafter that management recognized the cost benefit ration of their actions. Tactics at the airlines resulted in a 95% reduction in the use of "virgin halon" and saved the company $40,000 per year. Additional changes at the airline included modifications in the type of printed materials used by the company itself (i.e., switch to recycled goods) as well as changes in the amount of paper goods found on-board flights.

Source Reduction Clearinghouse [2,3]

Similar to the establishment of state funded grant programs designed to promote pollution prevention efforts, the Pollution Prevention Act of 1990 also called for the development of a *source reduction clearinghouse*. The underlying purpose of this program would be to serve as the crossroads of source reduction technology transfers. As such, the clearinghouse would be used as a means to collect data from state funded/matched source reduction programs. Specifically though, the responsibilities of the clearinghouse, as listed in the Act, would include:

1. To serve as the center for source reduction technology transfers
2. To mount active outreach and education programs by the states to further the adoption of source reduction technologies
3. Collect and compile information reported by states receiving grants, including success rates of each program

Source Reduction and Recycling Data Collection [2,3]

The final directive of the Pollution Prevention Act of 1990 mandates that every operator of a facility required to file an annual toxic chemical release form will also now be required to include a toxic chemical source reduction and recycling report for the preceding calendar year with their annual submission. This new initiative, designed to assist with pollution prevention methodologies, will include all chemicals required to be reported under previously detailed statutes. The extent of the source reduction effort to be instituted at each facility will depend on the following pieces of information:

1. The quantity of the chemical entering any waste stream prior to recycling, treatment or disposal
2. The amount of the chemical from the facility which is recycled during a calendar year as well as the percent change of recycled material from the preceding year and the process of recycling that was used

3. The source reduction practices used with respect to that chemical during a calendar year
4. The amount of chemical expected to be reported for the next two years.
5. A ratio of production of that chemical from the current year to the preceding year's net
6. The techniques that were used to identify source reduction opportunities. Examples may include employee recommendations, external and internal audits, material balance audits and alike
7. The amount of toxic chemical released into the environment which resulted in a catastrophic event, remedial action or alike
8. The amount of chemical from the facility which is treated at the facility during a calendar year and a percentage change from the current and previous years

Program Progression [3]
In an effort to ensure that all components of the Pollution Prevention Act were and continue to remain functional and intact, Congress attached a final section to the legislation that mandated the EPA to submit biennial reports containing detailed descriptions of actions taken to promote pollution prevention strategies. These reports are to contain the following bits of information:
1. An analysis of industry specific data collected under the Pollution Prevention Act of 1990
2. An analysis of the usefulness and validity of the data collected under the Act
3. Identification of regulatory and non-regulatory barriers to source reduction and of opportunities for using existing regulatory programs and incentives and disincentives to promote and assist source reduction
4. Identification of industries and pollutants that require priority assistance in multi-media source reduction
5. Recommendations as to incentives needed to encourage investment and research and development in pollution prevention methodologies
6. Identification of opportunities and development of priorities for research and development in pollution prevention
7. An evaluation of the cost and technical feasibility, by industry and processes, of source reduction opportunities and current activities and an identification of any industries for which there are significant barriers to source reduction
8. An evaluation of the methods of coordinating, streamlining and improving public access to data collected
9. An evaluation of data gaps and data duplication with respect to data collected under the Federal environmental statutes

Supplementary State Program [2]
In addition to state grants that are mandated by the EPA, there also exists a host of other programs that serve the cause of source reduction technology and advancement.

PPITS

The Pollution Prevention Information Tracking System, (PPITS), is a "user friendly data base that houses the most up-to-date information on state grants." The system contains information on the Source Reduction and Recycling and Technical Assistance (SRRTA), Pollution Prevention Incentives for States (PPIS) and the RCRA Integrated Training and Technical Assistance (RITTA) grants. The data base is maintained by the EPA and is an invaluable resource for companies looking to obtain additional funds for pollution prevention initiatives at their workplace.

The establishment of an integrated information system has allowed PITTS to establish a series of program objectives. They include:

1. To build state pollution prevention capabilities
2. To test, at the state level, innovative pollution prevention approaches and methodologies
3. To foster coordination and exchange of information between federal agencies, state and local governments and the private sector
4. To target high-risk environmental problems in sectors but traditionally addressed by the EPA
5. To leverage EPA resources through seed money and well-targeted grants

National Roundtable of State Pollution Prevention Programs [2,4]

The National Roundtable of State Pollution Prevention Programs (NRSPPP) is a non-profit organization that provides "a national forum promoting the development, implementation, and evaluation of efforts to avoid, eliminate, or reduce waste or pollution generation." Members of this organization often include:

1. Federal, state, and local agencies
2. Businesses
3. Industry and trade organizations
4. Environmental organizations

Membership into the organization also has several associated benefits. For example, members have some influence over the construction of national environmental policy regarding source reduction as well as access to the most up-to-date network of state and local source reduction programs. Additional membership perks include access to information on current legislative developments, technical assistance, and related publications.

The Northeast (NE) States Pollution Prevention Roundtable established in 1989 was designed to improve the capabilities of state environmental officers in Connecticut, Maine, Massachusetts, New Hampshire, New York, and Vermont to institute successful pollution prevention initiatives. Objectives of the NEPPR include:

1. Managing a regional roundtable of state and local environmental programs
2. Establishing a resource library of information on source reduction

3. Conducting pollution prevention training sessions for state officials and industry representatives
4. Researching source reduction strategies and techniques
5. Coordinating joint policy and program development

Common Sense Initiative [2,4]

In an attempt to reinvent the country's environmental regulatory system the EPA established the Common Sense Initiative (CSI) in 1993. The foundation upon which this program was based included employing stakeholders as resources in finding more "flexible, cost-effective and environmentally protective solutions" for the war against pollution. These objectives were later dispersed amongst six defined sectors of the industry including – automobile manufacturing, computers and electronics, iron and steel, printing, metal finishing, and petroleum refining. Because the six industries comprise over 11% of the gross national product, employ over 4 million people, and account for over 12% of the toxins released by American industries, they serve as excellent pilot projects for testing and refining the CSI objectives and goals. Furthermore, associated benefits of industry involvement in CSI include leading to more operational flexibility, increasing industry responsiveness to market demands, and increasing the overall level of competition amongst varying companies.

Executive Orders [1]

In conjunction with the previously established Pollution Prevention Act of 1990, President Clinton created a series of executive orders in attempts to ensure source reduction initiatives. Such legislation was designed to require federal agencies to comply with pollution prevention tactics, recycling, and waste management strategies.

Executive Order 12856 [1]

Executive order 12856 was created in an attempt to ensure that all federal agencies conduct their facility management and acquisition activities, so that, to the maximum extent practical, the quantity of toxic chemical entering any waste stream, including release to the environment, is reduced through source reduction; that waste is recycled to the maximum extent possible; that any remaining wastes are stored, treated, or disposed of in a manner protective of the public health and the environment.

Executive Order 12873 [1]

Executive order 12873 was created to ensure that the head of each executive agency be required to incorporate waste prevention and recycling efforts into that agency's daily operations and work to increase and expand markets for recovered materials through greater federal government preferences and demand for such products. These agencies would also be expected to comply with executive branch policies for the acquisition and use of environmentally preferable products and services. Furthermore, each executive agency is mandated to have an established goal for solid waste prevention and recycling by 1995.

State and Local Regulations

In an effort to cater more specifically local needs, many states began instituting and implementing additional legislation in the name of pollution prevention methodology. While these regulations incorporated many of the same core concepts as the federal mandates, they often exceeded national policy by including regulations that were more stringent and comprehensive in detail. Examples of some of the polices implemented by state and local governments include the following:

1. Mandatory source separation
2. Disposal bans
3. Variable disposal rates
4. Anti-scavenging ordinances

The northeastern United States has some of the most progressive and stringent source reduction guidelines in the country. While such legislation is most likely a reflection of public wants and needs, it too reflects the dire straits, relative to landfill space, that these states are currently facing. [5] An example of one of these comprehensive types of pollution prevention legislation is found in Massachusetts' Toxic Use and Reduction Act (TURA).[1] Enacted in 1989, goals of this initiative includes a 50% reduction of toxic chemical generation or hazardous by-product production by industries between 1987 and 1997. Companies required to abide by TURA standards include those that employ the equivalent of at least 10 full time workers; conduct any of the business activities described by Standard Industrial Classification (SIC) Codes 10-14, 20-39, 40, 44-51, 72-72, 75-76; and is a large quantity generator.

Methods utilized for the attainment of such objectives included a multi-media approach to source reduction. As such, methodologies focused on pollution prevention strategies across air, water, and land. Substances subject to TURA jurisdiction include all of the substances on the EPA Section 313 Emergency Planning and Community Right-to-Know list and on the EPA Comprehensive Environmental Response, Compensation and Liability. [6]

New Jersey, like Massachusetts also passed local ordinances in attempts of catering more specifically to local needs. Under the New Jersey Pollution Prevention Act of 1990, facilities in certain industries were required to not only develop on-site pollution prevention plans, but also submit plan summaries and yearly progress reports to the New Jersey Department of Environmental Protection. [7] Interviews with several of these targeted companies revealed that a majority of the facilities found planning to be "worthwhile" in that the benefits often times extended beyond reduction goals. Likewise, many companies also reported having set higher reduction goals than the statutes actually mandate.[7]

Industry Involvement [4,8]

In an attempt to abide by the regulations and statutes of both the federal and local governments, many industries and companies have taken it upon themselves to initiate their own pollution prevention projects. One such example is the *Automotive Mercury Switch Collection and Recycling Project*, where pollution prevention efforts focus on the collection and recycling of mercury switches from the hoods and trunks of automobiles. Projections of the study estimate that through such source reduction efforts, 500 pounds of mercury will escape leakage into the Great Lakes Basin. Another example of industry innovation is the *Strategic Goals Program (*SGP. The SGP is a cooperative effort between New York State, the EPA, and the American Electroplaters and Surface Finishers Society (AESF), the National Association of Metal Finishers (NAMF), and the Metal Finishing Suppliers Association (MFSA). Objectives of this collaborative effort include testing new ideas that are both "bold and common sense" in nature for improving environmental protection by the metal finishing industry.

A final initiative that has also been designed by industry is the Environmental Accounting Project (EAP). Established in 1992, the mission of the EAPt is to "encourage and motivate businesses to understand the full spectrum of their environmental costs and integrate these costs into decision making." The foundation upon which the project was designed stemmed from concerns from outside stakeholders who felt that source reduction would not be adopted as the primary choice of environmental management unless cost benefit ratios were clearly observable. As a result of such concern, the EPA constructed a focus group of experts to assess the role that the EPA could have in such decision making. The product of this panel yielded the recommendations that centered on the establishment of a project (the EAP) that would assist businesses in understanding the benefits of source reduction tactics.

Conclusion

With the passage of the Pollution Prevention Act of 1990 and Executive Orders 12856 and 12873, the U.S. began its movement toward long-term environmental protection and conservation efforts. While source reduction policies have been implemented on several different levels (both nationally and locally), only 15 states have currently adopted ordinances that cater more specifically to the wants and needs of their communities. Involvement by industries in the war against pollution has, on the other hand, been enhanced greatly by company and industry projects that have a their crux means to illustrate the benefits of source reduction on a multitude of levels.

REFERENCES

1. **Cooke, S. and Berry, D.,** of Goodwin, Provtor and Hoar, *Massachusetts Environmental Law Handbook,* 2nd Edition, Govt. Institutes, Inc. July, 1997.

2. **USEPA**, Office of Pollution Prevention and Toxics, *Pollution Prevention: Pollution Prevention Incentives for States*, <http://www.epa.gov/p2>, Updated Jan. 31, 2000.

3. **USEPA**, Office of Pollution Prevention and Toxics, *Pollution Prevention*, <http://www.epa.gov>, Updated Jan. 31, 2000.

4. **USEPA**, Office of Pollution Prevention and Toxics, *About OPPT*, <http://www.epa.gov>, Updated Jan. 31, 2000.

5. **USEPA**, Office of Pollution Prevention and Toxics, *United States Code: Title 42- The Public Health and Welfare; Chapter 133- Pollution Prevention*, <http:www.epa.gov>, Updated Jan. 31, 2000.

6. **NEWMOA**, Northeast Waste Management Officials' Association, *Pollution Prevention Program*, <http://www.newmoa.org>, Updated 1999.

7. **MADEP**, Massachusetts Department of Environmental Protection Bureau of Waste Prevention, *Toxic Use Reduction*, <http://www.state.ma.us/dep>, Updated Feb. 17, 1998.

8. **NJDEP OPPPC**, Pollution Prevention Planning Report, *Evaluation of the Effectiveness of Pollution Prevention Planning in New Jersey*, <http://www.state.nj.us/dep>, Updated March 29, 2000.

9. **NYSDEC**, New York State Department of Environmental Conservation, *Pollution Prevention Projects*, <http://www.dec.state.ny.us>, Updated April 5, 1999.

10. **USEPA**, Environmental Accounting Project, *EA Project*, <http://www.epa.gov>, Updated April 14, 1998.

WEB LINKS FOR THIS CHAPTER

The Pollution Prevention Act (PPA) Summary
http://www.epa.gov/region5/defs/html/ppa.htm

Full Text of the PPA
http://www.epa.gov/opptintr/p2home/uscode.htm

Office of Pollution Prevention and Toxics
http://www.epa.gov/opptintr/p2home

Fact Sheet for Form R Reporting
http://earth2.epa.gov/techinfo/facts/epa/reqir-fs.html

Design for the Environment
http://www.epa.gov/opptintr/dfe

Massachusetts Toxic Use Reduction Act (TURA)
http://www.state.ma.us/legis/laws/mgl/21I-3.htm

Massachusetts TURA Publications and Forms
http://www.magnet.state.ma.us/dep/bwp/dhm/tura/turapubs.htm

Massachusetts Bureau of Waste Prevention Program
http://www.magnet.state.ma.us/dep/bwp/dswm/dswmpubs.htm

IDENTIFICATION and LABELING OF TOXIC SUBSTANCES

7

Marc Nascarella

Modern civilizations are virtually dependent upon the manufacture and use of thousands of unique chemicals. Many of those chemicals, if used improperly, have the potential for catastrophic effects on human health and the environment. The Toxic Substance Control Act Inventory contains over 70,000 chemicals that the EPA monitors for the public health impact.[1] Laws that ensure the safety and health of the public are drafted to address the potential of a chemical to cause bodily injury. A chemical may cause bodily injury by means of its use, misuse, transportation, disposal, or treatment. Therefore, the laws that are implemented address those specific areas.

STATE REGULATIONS

Some states have elected to enforce toxic substance laws similar to federal regulations. State regulations and the federal standard are alike in many respects. Both sets of regulations seek to improve the health and welfare of the general worker and resident through effective communication of toxic hazards.

There are, however, many unique differences between the two requirements. For example, the Massachusetts based standards are broader in terms of application to industries in the commonwealth of Massachusetts. Both the federal and state regulations apply to chemical manufacturers, importers, and distributors. However, the Massachusetts regulations apply to all employers in the state who use hazardous chemicals, while the federal standard applies only to manufacturers, defined as businesses in the standard industrial Classification Codes 20 – 39. The Massachusetts standards also allow for more community access to information under certain circumstances, which the federal regulations do not address.[2]

Another unique difference is the spirit of the regulations. The state regulations tend to make specific details on how to comply with the requirements. Whereas the federal standard makes goals and allows the employer to decide how to meet them. For example, the federal government requires that labels be legible, while the state regulations specify exactly what size type will be used.

FEDERAL TOXIC SUBSTANCES CONTROL ACT

The Federal Toxic Substances Control Act (TSCA 15 U.S. Code 2601) was enacted by Congress on October 11, 1976. This Public Law (94-469) was enacted to "*regulate commerce and protect human health and the environment by requiring testing and neccesary use restrictions on certain chemical substances, and for other purposes*". Recognizing in 1976 that insufficient data existed with respect to the effects of chemical substances and mixtures on human health, Congress authorized the administrator of the EPA to track the 70,000 industrial chemicals currently produced or imported in the U.S. An extract of the TSCA Inventory has been made available by Cornell University <*http://www4.law.cornell.edu/uscode/15/ch53.html*>. A searchable CD-Rom version of the TSCA Inventory, which includes SARA Title III data, is available through National Toxicology Information Service, at 703/487-4650 or through the NTIS web site:<*http://www.ntis.gov/index.html*>.

The EPA is charged with the ultimate administration of the TSCA. Imbedded in the responsibility is the requirement that the EPA perform a wide array of duties. These include but are not limited to the promulgation of rules and regulations, the review of premarket notifications and tests, the regulation of chemical substances and mixtures, the compilation of lists, the administration of grants, and the submission of reports to Congress. These duties are mandatory in nature and the EPA is subjcct to citizen suits if they fail to perform there duties.[3]

Prior to enactment of this landmark legislation, the federal legislation of chemicals was extremely limited and covered only small groups of chemical with virtually no control over new chemicals. TSCA provides specific guidance and directives that are not otherwise covered under the nation's other toxic substance acts such as the Federal Insecticide, Fungicide, and Rodenticide Act, and the Food Drug and Cosmetic Act (see Table 7-1).

For more information contact the Toxic Substances Control Act (TSCA) Assistance Information Service (202-554-1404), also known as the TSCA Hotline. You may also wish to contact the U.S. Government Printing Office (GPO). The GPO sells approximately 12,000 different printed and electronic publications that originate in various government agencies, including EPA. They also provide free online access to more than 70 databases of Federal Government publications, including the Congressional Record and the Federal Register. GPO's online information service, GPO Access, may be reached at <*http://www.access.gpo.gov*>.

The EPA repeatedly screens and evaluates new chemicals and mandates reporting or testing of those that pose a risk to human health or the environment. The EPA also has the authority to place restrictions on the production and use of materials that pose an "imminent risk." This portion of the act is termed the imminent hazard provision.

1. Firearms and ammunition subject to taxes under section 4181 of the Internal Revenue Code, 26 USC 4181
2. Food, food additives, drugs, cosmetics, or devices regulated by the Federal Food Drug and Cosmetic Act, 21 USC 301
3. Meat and meat products regulated by the Federal Meat Inspection Act, 21 USC 601
4. Eggs and egg products regulated by the Egg Products Inspection Act, 21 USC 1031
5. Poultry and poultry products regulated by the Poultry Products Inspection Act, 21 USC 451
6. Any pesticide regulated by the Federal Insecticide, Fungicide, and Rodenticide Act, 7 USC 136
7. Tobacco or any tobacco product
8. Any source material, special nuclear material, or byproduct material regulated by the Atomic Energy Act, 42 USC 2201

Table 7-1. Items excluded from coverage under TSCA[3]

Under this provision if the EPA administrator determines that it is likely that a substance pose an unreasonable risk to the health or the environment, he or she may impose a ban or limit the manufacture, processing, or distribution of a chemical substance after its publication in the *Federal Register*. The EPA administrator may also file an action in the U.S. district court against any imminently hazardous substance or mixture. For example, the EPA has banned the use of both asbestos and PCBs due to the human health hazard associated with both substances. The decision to ban both of these substances is based on human health data that is discussed in the following section.

PCBs

Polychlorinated biphenyls (PCBs) are a class of synthetic organic chemicals that contain 209 individual chlorinated chemicals (known as congeners). PCBs may be either oily liquids or solids of a light yellow or colorless appearance. They are both odorless and tasteless. PCB mixtures may also be referred to in the United States by their industrial trade name, Aroclor.[4]

PCBs resist burning and are thus good insulating material. They have seen extensive use as coolants and lubricants in transformers, capacitors, and other electrical equipment. The U.S. manufacture of PCBs was ended in 1977 because of increasing reports of environmental persistance and harmful human health effects. Some products that originally contained PCBs, and provided routes of exposure are old fluorescent lighting fixtures, electrical appliances containing PCB capacitors, old microscope oil, and hydraulic fluids.[4]

Prior to the production hault in 1977, PCBs were released in the air, water, and soil during their manufacture and use. Currently, PCBs may be released into the environment from hazardous waste sites that contain PCBs, improper disposal of PCB wastes, and leaks from electrical transformers containing PCBs. PCBs are persistant and may be carried long distances in the air where they reamin for approximately 10 days. In water the small amounts of the PCBs that may be released are accumulated, and are stored in sediments. These sediment PCBs build up in fish and marine mammals (termed bioaccumulation) and can reach concentrations in the food chain thousands of times higher than the levels in the water or sediments.

Exposure to PCB aerosols may cause irritation of the nose and lungs, and skin irritations, such as acne and rashes. It is not documented whether PCBs may cause birth defects or reproductive problems in humans. Some research have shown that children born to women who consumed PCB-contaminated fish had damage to their nervous systems at birth. However, it is not known whether the damage is directly linked to PCBs or other factors.

Animals that breath very high levels of PCBs experienced liver and kidney damage, while animals that consumed food with large amounts of PCBs had mild liver damage. Animals that consumed food with low levels of PCBs had liver, stomach, and thyroid gland injuries, and anemia, acne, and problems with their reproductive systems. Dermal exposure to PCBs in animals resulted in liver, kidney, and skin damage.[5]

The EPA has set a maximum contaminant level of 0.0005 milligrams PCBs per liter of drinking water (0.0005 mg/L). The EPA requires that spills or accidental releases into the environment of 1 pound or more of PCBs be reported to the EPA. The Food and Drug Administration (FDA) requires that milk, eggs, other dairy products, poultry fat, fish, shellfish, and infant foods contain not more that 0.2–3 parts of PCBs per million parts (0.2–3 ppm) of food.[4]

Implementing regulations, are found in 40 CFR Part 761. Some key provisions of the regulations and EPA policies are listed in Table 7-2.

It is has yet to be elucidated whether PCBs cause cancer in humans. In a long-term (365 days or longer) study, PCBs caused cancer of the liver in rats that consumed PCB mixtures. The Department of Health and Human Services (DHHS) has determined that PCBs may reasonably be anticipated to be carcinogens.[5]

For more information the EPA Office of Pollution Prevention and Toxin Homepage provides a comprehensive web site listing the effects of PCBs (both carcinogens and noncarcinogenic PCBs):
<http://www.epa.gov/opptintr/pcb/effect.htm>.

Asbestos

Asbestos is a naturally occurring group of minerals. It is mined in operations that resemble the mining of common materials such as such as iron, lead, and copper. The chemical structure of asbestos contains silicon, oxygen, hydrogen, and various other positively charged metal ions. Common forms of asbestos include chrysotile, amosite, and crocidolite. Chrysotile fibers are known for being the most pliable and cylindrical, and occur in bundles. Amosite and crocidolite fibers are best described as tiny razor sharp needles. The first commercial asbestos mining operation (chrysotile) was begun in 1870 at Quebec, Canada. Crocidolite was later mined. Asbestos was first mined in 1980 in South Africa. Amosite asbestos also from Africa and was initially mined in

1. Visual inspection and recordkeeping for PCBs
2. Disposal restrictions on use and burning of used oil containing PCBs
3. Spill prevention
4. PCB spill cleanup policy
5. Food and feed restrictions
6. PCB transformer fire regulations
7. Substitute dielectric fluid
8. Storage container specifications
9. Notification and manifesting rule
10. EPA policy on physical separation of PCBs
11. Reclassification of transformers
12. PCB fluorescent light ballast disposal
13. PCBs in laboratories

Table 7-2. Key provisions of 40 CFR Part 761

1916. Asbestos is a unique mineral due to the fact that it will turn into dust particles when crushed and will fracture into ultrafine fibers that may only be observed with a microscope. Often asbestos fibers are processed with a binding agent that holds them together, producing an asbestos containing material (ACM) that is both flexible and fire resistant. However, if the structural integrity of the asbestos containing material is compromised the asbestos may become airborne, and when inhaled may remain in the lungs for a long period of time. This will present a potential risk for severe health problems with a possible long latency period.[6] For more information on the health effects of asbestos refer to the following EPA web site: <*http://www.epa.gov/oar*> .

Individuals exposed to asbestos develop may or may not develop asbestos-related health problems. Inhaled, asbestos fibers have the ability to penetrate body tissues. They are often times permanenly deposited in the airways and lung tissue. Because asbestos fibers remain in the body, each exposure increases the risk of an asbestos-related disease. The EPA recommends a medical examination that includes a medical history, a breathing capacity test and a chest x-ray which may detect problems early. Currently there is no "safe" or threshold level for exposure to aerosolized asbestos. Both dermal contact and ingestion pose uncertain human health risks.[6]

Illnesses associated with asbestos exposure include asbestosis, lung cancer, mesothelioma, and cancers in the esophagus, larynx, oral cavity, stomach, colon, and kidney. Asbestosis is defined as a serious, chronic, non-cancerous respiratory disease. Inhaled asbestos fibers aggravate lung tissues, which promotes scarring of the tissue. Common symptoms of asbestosis include shortness of breath, and a dry crackling sound in the lungs while inhaling. In later, more advanced stages the disease may cause cardiac failure. Lung cancer is considered the number one cause of deaths due to asbestos exposure. Common symptoms of lung cancer are coughing and a marked breathing change. Lung cancer incidence is greatly increased by the presence of both cigarette smoking and asbestos exposure. Mesothelioma is a rare form of cancer common to the thin membrane lining of the lungs, chest, abdomen, and sometimes the heart. Virtually all of the 200 cases of mesothelioma reported in the U.S. each year are linked with asbestos exposure. According to the EPA approximately 2% of all miners and textile workers who work with asbestos, and 10% of all workers who were involved in the manufacture of asbestos-containing gas masks, contract mesothelioma. Cancers in the esophagus, larynx, oral cavity, stomach, colon, and kidney may be caused by ingesting asbestos. [6,7]

More that 3000 products in use today contain asbestos. Most of these are materials are used in heat and acoustic insulation, fire proofing, and roofing and flooring. Some of the more common products that may contain asbestos are listed in Table 7-3.

Radon

Radon is a naturally occurring gas that seeps out of rocks and soil. Radon originates from the element uranium. The rate of radon seepage varies by geographic region, partly because the amounts of uranium in the soil are considerably different. Radon escapes

1. Pipe and duct insulation
2. Building insulation
3. Wall and ceiling panel
4. Carpet underlays
5. Roofing materials
6. Artificial fireplaces and materials.
7. Patching and spackling compounds
8. Brake pads and linings
9. Pot holders and ironing board pads
10. Hair dryers
11. Floor tiles
12. Electrical wires
13. Textured paints
14. Cements
15. Toasters and other household appliances
16. Furnaces and other furnace door gaskets

Table 7-3. Asbestos Containing Products [6]

from the ground into outdoor air and also into the indoor air environment of homes via underground levels (basements).[8] Outdoor radon is not viewed as a problem due to the high rate of mixing with cleaner radon free ambient air. In contrast, among indoor environments radon concentrations can exceed safe levels and may pose a serious health hazard.

Radon is chemically inert and electrically uncharged, however, it is radioactive, meaning that radon atoms in the air may decay into different atomic structures. When the new atom is formed via decay (called radon progeny) they are electrically charged and may bind to tiny dust particles in indoor air. These particles represent an inhalation hazard by adhering to the lining of the lung. The inhaled particles decay and emit alpha radiation, which may damage cells in the lung. Alpha radiation may alter DNA of lung cells. This DNA damage is a step in the propagation of cancer in human cells. The radiation exposure is thought to be contained in the lung, and therefore lung cancer is the main hazard associated with radon exposure.[8]

On October 28, 1988, Congress created Title III of TSCA to address these problems and create a national long-term goal of reducing the radon levels in buildings to ambient levels. Title III directs EPA to develop model construction standards and techniques for controlling radon levels within new buildings. These standards are to incorporate geographic differences in construction types and materials, geology, weather, and other variables that affect radon levels in buildings. Title III also directs EPA to assist individual state radon abatement programs by providing technical assistance, education programs, and information on mitigation alternatives and measurements.

Title III requires EPA to conduct a study to determine the extent of radon contamination in the nation's school buildings and in federal buildings. In each case, EPA is to identify "high-risk" areas for attention, based on geological data, data on high radon levels in buildings near schools or federal sites, and physical characteristics of school or federal buildings. For a comprehensive listing of Web sites related to radon refer to the following link <*http://www.radon.com/radon/radon_links.html*>.

The most recent information available to estimate the risks posed by exposure to radon in homes is the National Research Council study, which has been carried out by the Committee on Biological Effects of Ionizing Radiation (BEIR) VI. [9] The BEIR report is considered to be the most comprehensive and scientific review of the health effects of radon. The EPA has used the report in its evaluation and review of health based standards as well as summarized its contents for public disclosure (see Table 7.3). The BEIR report states that the most direct way to assess the risks posed by radon in homes is to measure radon exposures among people who have lung cancer relative to exposures among people who do not have lung cancer. The studies already conducted and those underway have yet to produce a definitive answer. This is due mainly because the risk is small at low exposures encountered from most homes, combined with the difficulty in estimating radon exposures that people have received over their lifetimes.

Given the inherent difficulty in residential exposures the BEIR VI committee chose to use the occupational exposure and lung-cancer information from miners, who are more exposed to higher levels of radon, and estimate from those data the risks posed by radon exposures in homes. The committee chose 11 major studies of underground miners, which together involved about 68,000 men, of whom 2700 have died from lung cancer, as the focus of their attention. A summary of the report may be seen in Table 7-4.

The report concluded that the risk of lung cancer from smoking is much greater than the risk of lung cancer from indoor radon. The report also found a synergistic (or multiplicative) effect of both smoking and lung cancer. However, the estimated 15,400 or 21,800 deaths attributed to radon in combination with cigarette smoking and radon alone in nonsmokers is clearly a serious public health problem.

For general information on radiation contact the office of Air and Radiation (OAR)

at the EPA <*http://www.epa.govoar/oarpubs.html*>. For radon specific information contact the EPA web site on radon at
<*http://www.epa.gov/iaq/radon/pubs*

FEDERAL HAZARDOUS SUBSTANCES ACT

According to the Federal Hazardous Substances Act (**HSA**) of 1960, household products are hazardous if they are one of the following:

1. toxic
2. corrosive
3. flammable
4. combustible
5. radioactive

---

1. BEIR VI confirms EPA's position for the last decade that radon is the second leading cause of lung cancer and a serious public health problem.
2. NAS has estimated that 12% of lung cancer deaths in this nation are linked to radon. NAS' best estimates are that radon causes in the range of 15,000 to 22,000 lung cancer deaths each year. EPA's best estimate has been 14,000 annual lung cancer deaths due to radon.
3. This report represents the most definitive accumulation of scientific data gathered on radon since the NAS' BEIR IV report of 1988. It affirms once more that EPA's radon policies are well grounded in strong science.
4. NAS further noted that the data from radon studies in homes is consistent with studies of radon health effects in mines and that lung cancer can be reduced by limiting exposure to radon in homes.
5. NAS concluded that, although there is some uncertainty in these esti-mates, lung cancer in the general population is an important public health concern.
6. The report concludes that many smokers will get lung cancer due to their radon exposure, which exacerbates the effects of smoking.
7. The BEIR VI report found that even very small exposures to radon can result in lung cancer. In fact, NAS concluded that no evidence exists that shows a threshold of exposure below which radon levels are harmless.
8. EPA is committed to protecting the public's health. EPA has in place a voluntary program to find the homes with high radon levels. We recom-mend that people do a simple home radon test and if high levels of radon are found, we recommend reducing those high levels with straight for-ward techniques.

Table 7-4. EPA's Summary of BEIR VI Radon Report*

*The summary of the BEIR VI report is available from the US EPA at <*http://www.epa.gov/ied-web00/radon/beirvi.html*> or the entire publication is availble from the National Academy of Science online at <**http://www.nationalacademies.org**>.

6. an irritant
7. a strong sensitizer
8. capable of generating pressure through decomposition, heat, or other means
9. capable of producing substantial injury through reasonably foreseeable handling or use, including ingestion by children

The Consumer Products Safety Commission (**CPSC**) is required under HSA to evaluate substances for these hazardous characteristics and to identify those substances that are hazardous. CPSC must include those substances already defined as hazardous by other programs, such as the Occupational Safety and Health Administration (OSHA) regulations and EPA regulations of hazardous wastes. HSA requires CPSC to follow certain notice and hearing procedures when considering whether to define a substance as hazardous. CPSC regulations prescribe tests for assessing whether products are hazardous, and how to accurately label them. Insecticides, foods, drugs and cosmetics; fuels used domestically; and radioactive materials are currently regulated by other laws and are specifically excluded from the labeling acts.[10]

Hazardous substances must be labeled with written, printed, or graphic matter appearing on the immediate and outside containers of the substance. Label statements must be prominently located, in English, and must provide specific information (see Table 7-5).

For more information refer to the following site supported by the EPA and Purdue University. The information contained therein provides general information on toxic substance identification, storage, and disposal: *<http://www.epa.gov/grtlakes/sea-home/housewaste/src/open.htm>*.

---

1. Common, usual or chemical name of the substance
2. Name and address of the manufacturer, packer, distributor or seller,
3. Affirmative statement of the principal hazards
4. First aid instructions
5. Handling and/or storage instructions if necessary
6. "Signal Words" such as DANGER, WARNING, or CAUTION, to highlight the substance's hazardous characteristics;
7. Directions adequate to protect children from the hazard.

Table 7-5. Information required on hazardous substance

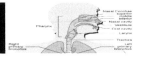

REFERENCES

1. **U.S. EPA**, Chemical Testing and Information Gathering, General Information April 2000. <http://www.epa.gov/opptintr/chemtest/index.htm>

2. **Inter/Face Associates**, *Employee Right-to-Know: a guide to Massachusetts and federal regulations,* Inter/face Associates, 62 Washington St., Middletown, CT, 1984.

3. **Druley, R. M. and Ordway, G.L.,** The Toxic Substances Control Act, Bureau of National Affairs Inc., Washington, D.C., 1981 (revised ed.) 410.

4. **U.S. EPA**, Office of Pollution Prevention and Toxins, Fibers and Organics Branch (7404), *Polychlorinated Biphenyls Publication.* U.S. EPA, Washington,D.C.

5. **ASTDR**, *Toxicological Profile for Selected PCB's,* U.S. Department of Health and Human Services, Atlanta, GA., 1993.

6. **U.S. EPA,** *Integrated Risk Information System (IRIS) on Asbestos, Natural.* Environmental Criteria and Assessment Office, Office of Research and Development, Cincinnati, OH., 1993.

7. **Calabrese, E.J. and Kenyon, E.M.,** *Air Toxics and Risk Assessment,* Lewis Publishers, Chelsea, MI., 1991.

8. **ATSDR**, *Toxicological Profile for Radon* (Draft), U.S. Department of Health and Human Services, Altanta, GA., 1990.

9. **Health Effects of Exposure to Radon**, BEIR VI, Committee on Health Risks of Exposure to Radon, National Academy Press, 1999, 516.

10. **Vance, B., Weinsoff, D.J., Henderson, M.A., and Elliot, J.F.**, *Toxics Program Commentary: Massachusetts*, Specialty Technical Publishers, Inc. North Vancouver, B.C. Canada, 1992.

WEB LINKS FOR THIS CHAPTER

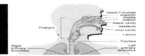

EPA web site on the effects of PCBs (both carcinogens and noncarcinogenic)
http://www.epa.gov/opptintr/pcb/effect.htm>

EPA web site on the health effects of asbestos
http://www.epa.gov/oar

GPO Access
http://www.access.gpo.gov

Full text of TSCA
http://www.law.cornell.edu/uscode/15/ch53.html

Massachusetts Hazardous Substances Labeling Law
http://www.magnet.state.ma.us/dep/matrix.htm

My Radon Coloring Book (a must for the kids!)
http://www.epa.gov/rgytgrnj/kids/mrcb1.htm

PCB Disposal Amendments
http://www.epa.gov/opperid1/icr/1729ss02.htm

SARA Title III data at NTIS web site
http://www.ntis.gov/index.html

TSCA Inventory
http://www.pdc.cornell.edu/issearch/tscasrch.html

TSCA Product Labeling Requirements
http://www.epa.gov/opptintr/environmental-labeling/docs/tsca.pdf

Websites related to radon
http://www.radon.com/radon/radon_links.htm

UNDERGROUND STORAGE TANKS

8

Julie Vander Ploeg

L eaking underground storage tanks (LUSTs) are recognized as a major environmental problem, which has prompted federal and state legislators to create standards for the construction, monitoring, and removal of the tanks.[1] Over one million UST systems in the U.S. contain petroleum or hazardous substances regulated by the U.S. Environmental Protection Agency.[2] Many of these USTs have leaked or are currently leaking. The EPA has reported, as of October 1996 almost 318,000 UST releases have been confirmed. The EPA also estimates that the total number of confirmed releases could reach 400,000 in the next few years.[3] Contamination occurs from releases due to careless maintenance and poor filling practices.[4] After a leak, USTs contaminate soil, ground water supplies, and endanger human health. As result, in 1984 Subtitle I was added to the Resource Conservation and Recovery Act (RCRA). Since then a tremendous amount of resources have been used to clean up the contamination from USTs. [5]

OBJECTIVES OF THE PROGRAM

The federal underground storage tank (UST) regulations are part of the Hazardous and Solid Waste Amendments of 1984, which fall under the Resource Conservation and Recovery Act. Although the regulations of USTs are attached to RCRA, it is important to note that the regulations fall into a separate section known as Subtitle I. [4]

More information is available at the folllowing sites:

http://www.epa.gov/reg5oopa/defs/html/cerclea.htm
http://www.epa.gov/osw
http://wwwclay.net:80
http://www.mlb.com/env.htm
http://www.smallbiz-enviroweb.org/lawepalinks.asp

Subtitle I of RCRA requires the Environmental Protection Agency (EPA) to develop a comprehensive regulatory program for tanks storing petroleum or certain hazardous substances. These substances do not include other hazardous substances, which are regulated in different sections of RCRA.[8]   The Comprehensive Environmental Response, Compensation, and Liability Act (CERCLA) Section 101(14), contains a list of hundreds of substances designated as "hazardous." These hazardous substances are the same substances that can be stored in USTs, except for those that are listed as hazardous wastes under RCRA Subtitle C.   More information is available through the EPA's RCRA/Superfund hotline 1-800-424-9346.

As a result, the EPA under direction from Congress published regulations that require the owners and operators of new and existing tanks to prevent, detect, and clean up releases.[5]

Subtitle I of RCRA was amended by Congress in 1986. As a result, the Leaking Underground Storage Tank Trust Fund (**LUST**) was created.[5] The two purposes of the fund are:

1. To oversee cleanups by responsible parties.
2. To pay for the cleanups at sites where the owner or operator is unknown, unwilling, or unable to respond, or which require emergency action.

The amendments also required owners and operators to demonstrate that they are financially capable of cleaning up releases and compensating third parties for the resulting damages.[5]

FEDERAL REQUIREMENTS FOR UNDERGROUND STORAGE TANKS

Underground Storage Tank (UST) owners or operators must adhere to the following regulations in order to protect human health and the environment.[1]

1. If a UST was installed after December 22, 1988, the tank is required to meet requirements under new USTs. These requirements include leak detection, spill, over spill, and corrosion protection. Leak detection requirements are illustrated in Figure 8-1.
2. If the UST was installed before December 22, 1988 leak detection must be in place by December 1993 and spill, overflow, and corrosion protection need to be in order by December 1998.
3. Owners and operators must report a release once a leak of more than 25 gallons is confirmed. The leak must be reported to the proper authorities with in 24 hours.
4. Owners and operators must follow closure requirements. Tanks must be properly closed, either temporarily or permanently.
5. Financial responsibility belongs to the owner or operator. 40 CFR Part 280 lists the financial responsibilities associated with USTs. To simplify, the owner is responsible for cleanup costs of leaking tanks and to compensate others for bodily injury and property damage.

To meet these requirements, owners must upgrade, replace, or close their current system.[1] More information is available at the following site:
http://www.epa.gov/swerust1/pubs/index.html

What is an Underground Storage Tank?

The definition of an underground storage tank (UST) is a tank and any underground piping connected to the tank that has at least 10% of its combine volume underground. Only tanks containing petroleum and specified hazardous substances apply to federal regulations.[1]

There are certain tanks that are exempt from any federal regulations. Tanks that are exempt can be found in 40 CFR Part 280.2 The following tanks are exempt:

1. Farm and residential tanks of 1100 gallons or less capacity holding motor fuel used for noncommercial purposes
2. Tanks storing heating oil used on the premises where it is stored
3. Tanks on or above the floor of underground areas, such as basements or tunnels

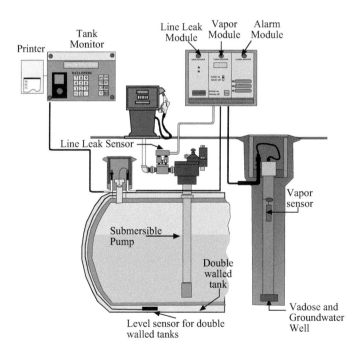

Figure 8.1 Leak detection methods. Source: U.S. Environmental Protection Agency.[1]

4. Septic tanks and systems for collecting storm water and
    wastewater
5. Flow-through process tanks
6. Emergency spill and overfill tanks

Upgrading USTs

New underground storage tank (UST) systems are those systems that were installed after December 22, 1988. Owners and operators are required to meet the following *before* installation:[6] See Table 8.1 for effective dates for upgrading and Table 8.2 for requirements.

1. Leak detection
2. Spill and overfill protection
3. Corrosion protection

Existing UST systems need the same requirements by December 22, 1998. However, to meet the following requirements the tank must be upgraded. Existing UST systems needed to have the following requirements:[7]

1. Leak detection by December, 1993
2. Spill and overfill protection by December 22, 1998
3. Corrosion protection by December 22, 1998

Leak Detection

All underground storage tanks must have leak detection. Owners and operators are required to use one of the following monthly monitoring leak detection methods.[6]

1. **Secondary containment and interstitial monitoring:** Secondary contain-

When Do You Have To Act?

Type of Tank and Piping	Leak Detection	Spill and Overfill Protection	Corrosion Protection
New Tanks and Piping (Installed after Dec. 22, 1988)	At Installation	At Installation (does not apply to piping)	At Installation
Existing Tanks & Piping (Installed before Dec. 22, 1988)	No later than Dec. 1993	No later than Dec. 22, 1998 (does not apply to piping)	No later than Dec. 22, 1998

Table 8.1 Effective Dates for all Upgrades. Source: U.S. Environmental Protection Agency[1] For more information go to <**http://www.epa.gov/swerust1/ pubs/ index.html** >

What Do You Have To Do?

Leak Detection	
New Tanks	Monthly monitoring; or Inventory control plus tank tightness testing** (only for 10 years after installation)
Existing Tanks	Monthly monitoring; or Inventory control plus tank tightness testing (only for 10 years after spill, overfill and corrosion protection); or Inventory control plusa nnual tank tightness testing (only until December, 1998).
New and Existing Pressurized Piping	Automatic shutoff device or flow restrictor or continuous alarm system; AND Annual line tightness test; or Monthly monitoring* (except automatic tank gauging)
New and Existing Suction Piping	Monthly monitoring;* or Line tighness testing every three years; or No requirements (if system has the characteristics described on page 11)*
Spill & Overfill Protection	
All Tanks	Catchment basins; AND Automatic shutoff devices or overfill alarms or ball float valves
Corrosion Protection	
New Tanks and Piping	Coated and cathodically protected steel; or Fiberglass reinforced plastic (FR); or Steel tank clad with FRP (does not apply to piping)
Existing Tanks and Piping	Same options as for new tanks & piping; or Cathodically protected steel; or Tank interior lining; or Tank interior lining AND cathodic protection

Table 8-2 EPA requirements for leak detection, spill, overfill, and corrosion protection. Source: U.S. Environmental Protection Agency[1] For more information see *http://www.epa.gov/swerust1/pubs/index.html.*

*   Monthly Monitoring includes: Interstitial Monitoring; Automatic Tank Gauging; Vapor Monitoring, Groundwater Monitoring; Statistical Inventory Reconciliation; and other methods approved by the regulatory authority.

**   Tanks 2000 gallons and smaller may be able to use manual tank gauging.

ment must be installed around the UST; this can be a vault, liner, or double-walled structure. Leaks are detected by using interstitial monitoring. New USTs are required to use this method.

2. **Automatic tank gauging systems (ATG):** These are monitors that provide information on temperature and product level. The monitor can calculate changes in volume, which can detect a leak.

3. **Vapor monitoring**: Vapor monitors can measure vapor in the soil around the tank and piping to determine the presence of a leak.

4. **Groundwater monitoring:** This monitoring device can detect the presence of liquid product floating on groundwater.

5. **Statistical inventory reconciliation (SIR):** SIR is a type of computer software that conducts statistical analysis of inventory, delivery, and dispensing data collected.

6. **Manual tank gauging:** This method can only be used on tanks that are 2000 gallons or smaller. Manual tank gauging does not work on larger tanks or on piping. This method requires the tank to be out of operation for 36 hours to measure the contents of the tank. Manual gauging requires the owner to physically measure the tank. Placing a measuring device similar to an oversized meter stick into the tank and recording the measurement does this. This measurement is the amount of fluid remaining in the tank.

7. **Tank tightness testing and inventory control:** This is a combination of two methods. Tank tightness is a test that checks the soundness and pressure of the tank. Inventory control testing requires the owner or operator to take daily measurements of the tank's contents and monthly calculations to prove that the tank is not leaking.

Spill Protection

More information is available at: *<http://www.epa.gov/swerust1/pubs/index.html>*. Most spills occur from human error. For example, spills often occur at the fill pipe when the delivery truck's hose is disconnected. These mistakes can be eliminated by following standard filling practices. If standard filling practices such as making sure there is room in the tank for delivery and the driver watches the delivery at all times, then nearly all spills can be prevented.

A solution to spill protection is catchment basins. A catchment basin is used to protect against spills. The basin is a bucket that is sealed around the fill pipe. Basins range in size from those having the ability to hold a few gallons to holding much larger quantities. The larger the basin the more protection is available.[1]

Overfill Protection

Overfills occur when the tank is overfilled. Overfills release much larger volumes than spills at the fill pipe and through loose fittings on the top of the tank, or a loose vent pipe.   If the tank was filled to its appropriate level the tightness of these fittings would have not been a problem.[6] In order to prevent overfills the EPA recommends the following.[6]

1. Make sure enough room is in the tank for a delivery.
2. Watch the delivery to prevent overfills and spills.
3. Use equipment to protect against overfills.

Automatic Shutoff Devices- A fill pipe is operated by a float mechanism that will cause the pump to shut off when the tank is full. This is important because if the driver is not paying attention the value will close and no more liquid can be delivered into the tank.

Overfill Alarms- Overfill alarms will be activated when the tank is 90% full or within one minute of the tank being overfilled. The alarm provides efficient time for the driver to shut off the valve. However, the alarm will only work if the driver is alert and is able to respond quickly. [6]

Ball Float Valves- Ball floats are placed in the vent line and rise with the input of product. The float will restrict vapor flowing out the vent line and create enough backpressure to restrict product flow into the tank. However, the float is only efficient if the fittings on the tank are tight, if they are loose this technique is not recommended. [6]

Corrosion Protection

Corrosion happens when bare metal, soil, and moisture conditions create an underground electric current that destroys hard metal. As time passes, corrosion can create holes, which will result in leaks.[6] There are certain tanks that are exempt from corrosion protection:

1. The tank and piping are made completely out of corrosion resistant material (fiberglass).
2. The tank and piping are made out of steel having corrosion-resistant coating and having cathodic protection.
3. The tank is made of steel clad with a thick layer of corrosion resistant material.

Corrosion protection can easily be overcome by one of the following:[8]

1. Cathodic protection; (Figure 8.2)
2. Interior lining; or
3. A combination of both

Cathodic protection- There are two types of cathodic protection: impressed current and sacrificial anode systems. Both protect the UST because corrosion-causing currents are diverted away from the tank.[8]

Interior lining- This is a thick layer of corrosion resistant material, which is used to line the interior of the tank.[6]

Combination of cathodic protection and interior lining- This is a combination of the two previous solutions.[6]

Underground Storage Tank Closure
More Information is available at:<***http://www.epa.gov/swerust1/pubs/index.html***>

To prevent unnecessary releases into the environment, the EPA has set up requirements for the correct closure of USTs. There are two different types of closure, temporary closure and permanent closure. Temporary closure allows owners and operators to temporally close their tanks for up to 12 months. However, owners and operators must continue leak detection and corrosion methods. Also if the tank is closed more than 3 months, vent lines must remain open, but other lines, pumps, manways, and ancillary equipment must be capped and secured. [6]

After 12 months of closure, owners and operators have three options:

1. Tanks must be closed permanently
2. Ask the regulatory authority for an extension beyond 12 months if contamination is present on your site
3. UST can remain closed if the tank meets the requirements for new or upgraded USTs and requirements for temporary closing

If the operator or owner decides to permanently close their UST, regulatory authority must be notified at least 30 days before the tank is closed. Site contamination must also be determined. If there is contamination, corrective action must be taken. Finally, the UST must be removed from the ground or properly left in the ground. In both cases the tank must be emptied and cleaned of all dangerous vapors, liquids and sludge. [9]

Record-Keeping
More Information is available at <***http://www.epa.gov/swerust1/pubs/index.html***> Record keeping begins once a UST is installed and continues until the tank is permanently closed. Even though the tank may be permanently closed records still need to be kept. There are several requirements associated with recorded keeping, including installation notification, release notification, and closing notification.

Installation notification-When the UST is installed a notification form must be filled out. This form is available through the state in which the tank is being installed. The purpose of this form is to provide information on the UST and information on certification of correct installation.[9]

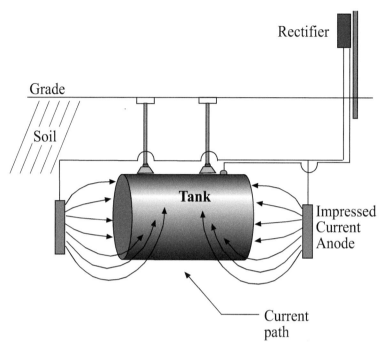

Figure 8.2. **Impressed Current System**. Source: U.S. Environmental Protection Agency.[1]

Release notification-Suspected leaks must be reported to the proper regulatory authorities. If the release is confirmed, records also must be kept on corrective action taken or plan to be taken to correct the damage caused by the UST.356

Closing notification-The proper regulatory authority must be notified, 30 days before the UST is permanently closed.[6]

Finally, it is important that owners and operators of UST's check with their state and local requirements for record keeping, because reporting requirements may vary in different areas. If you are not sure of your regulatory authority, there is a listing of federal and state agencies in the appendices.[6]

Along with the proceeding requirements owners and operators must keep records that can be provided to an inspector during an on site visit to prove that the facility is in order381. If there is ever a doubt of what records to keep and which to get rid of, remember when ever in doubt, keep it.

Records on leak detection and performance must be kept. These records should include last years monitoring results, performance claims from the leak detection manufactures, and records on repairs or adjustments made to any equipment391.

Records on corrosion protection must also be kept. These records must include inspections and test of the protection system. Records must also be kept on any repairs or upgrading done on the entire UST system. Records must also be kept on the closing procedures of the UST. These records should be kept for at least three years. The final record that must be kept is on the owners/operators financial responsibility.[1]

Release Reporting

If a release is suspected, the owner or operator at the facility must report the suspected leak to the proper regulatory authorities. A list of regulatory authorities is listed in the appendices. Through tightness testing the suspected leak needs to be determined if it is an actual leak. If there is an actual leak immediate action must be taken to fix the leak, followed by cleanup action. [1]

CLEANUP PROGRAM

There are more than one million underground storage tanks in the United States. EPA estimates that in the next several years, 400,000 confirmed releases will be reported. Most of these releases are discovered during the closure or replacing of USTs.

The EPA established a cleanup program under Subtitle I of the Resource and Recovery Act (RCRA). Under Subtitle I owners and operators are required to:

1. Report a release
2. Removing its source
3. Decrease fire and safety hazards
4. Investigate the extent of contamination
5. Clean up soil and groundwater

Under this program the EPA is trying to help state and local governments, make cleanups faster, cheaper, and more effective. The three following are solutions the EPA implemented.[1]

Risk-based decision making- This is used to consider the potential risks associated with releases and use this knowledge to make decisions about corrective actions. Risked-based decision making is encourage by the EPA because current and potential risks of USTs can be used to focus onsite assessment, classify sites, and determine if further action is necessary for cleanup.[1]

Alternative cleanup technologies- The EPA is working with contractors, consultants, tank owners, and states to encourage alternatives for site assessment and cleanup. To do this, the EPA is providing training, demonstrations, and outreach projects.[3]

Streamlining- Through the help of EPA staff and consultants, states are taught Total Quality Management techniques to improve cleanups.[6]

STATES AND LOCAL ROLE

EPA gave states and local governments the authority to oversee the management of USTs, because of the following reasons: [5]

1. State and local governments are closer to the situation.
2. Subtitle I allows state UST programs approved by the EPA to operate instead of the federal program.
3. State programs tend to be more stringent in their regulations compared to the federal requirements.

However, before states can implement their programs the EPA must recognize the program and deem it appropriate to be implemented. First, the EPA must approve state programs. If the state meets the following criteria the program is approved.[5]

1. Standards set for performance criteria are no less stringent than federal standards
2. Provisions for adequate enforcement
3. Regulates at least the same UST's as the federal standards

FINANCIAL RESPONSIBILITY

Under Subtitle I of the Resource and Recovery Act (RCRA), financial requirements were designed to make sure someone could pay the costs of cleaning up leaks and compensating third parties.[5] The owner and operator have all financial responsibility. If the owner and operator are two different individuals then a decision needs to be made between them of whom will be responsible.[6]

As a result of financial responsibility, the Leaking Underground Storage Tank (LUST) Trust Fund was initiated. This fund was established by Congress under Subtitle I of RCRA to provide money for cleanups for where the owner is unwilling, unknown,or unable to respond, and money for corrective action taken by the owner or operator.[6] Congress set up this program because they realized cleanups can be very costly. The fund is financed by a 0.1 cent tax on each gallon of motor fuel sold in the U.S. The program went into effect on December 31, 1995. As of April 1996, about $1.64 billion had been collected. Congress had given $595 million to the EPA in 1996. About $510 million went to state programs for administration, oversight, and cleanup work (LUST).

In order for the fund to work and for a state to receive money they must have an agreement with the EPA to use the money in a manner in which it was intended. The money is divided among EPA regional offices and additional money is given based on the number of releases, the number of notified petroleum tanks, the number of residents relying on groundwater for drinking water, and the number of cleanups.[10]

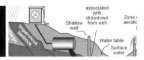

REFERENCES

1. **U.S. EPA**, *Musts for USTs: A Summary of Federal Regulations for Underground Storage Tank Systems,* Government Printing Office, Washington, D.C., 1995.

2   **U.S. EPA**, *Dollars and Sense: Financial Responsibility Requirements for Underground Storage Tanks*, Government Printing Office, Washington, D.C., 1995.

3. **U.S. EPA**, *Cleaning up Releases*, April 15, 1998. <http://www.epa.gov/swerust1/overview.htm>

4. **Thomas. A.**, et al., *Environmental Law Handbook* 15th ed.,Government Institutes, Maryland, 1999.

5. **U.S. EPA**, *Overview of the Federal UST Programs,* April 15, 1998. <http://www.epa.gov/swerust1/overview.htm>

6. **U.S. EPA**, *Don't Wait Until 1998: Spill, Overfill, and Corrosion Protection for Underground Storage Tanks*, Government Printing Office, Washington, D.C., 1994
.
7. **U.S. EPA**, *Are you upgrading an Underground Storage Tank System?,* Government Printing Office, Washington, D.C., 1997.

8. **U.S. EPA**, *Preventing releases*, April 15, 1998. <http://www.epa.gov/swerust1/fsprevent.htm>

9. **U.S. EPA**, *Closing Underground Storage Tanks*, Government Printing Office, Washington, D.C., 1996.

10. **U.S. EPA,** *Leaking Underground Storage Tank Trust Fund,* April 15, 1998. <http://www.epa.gov/swerust1/1tffacts.htm>

WEB LINKS FOR THIS CHAPTER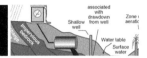

EPA Office of Underground Storage Tanks Homepage
http://www.epa.gov/swerust1/pubs/index.htm

Upgrading of Existing USTs
http://www.epa.gov/swerust1/pubs/upgrade.pdf

Closing USTs
http://www.epa.gov/swerust1/pubs/clo.pdf

UST Leak Detection Evaluations
http://www.epa.gov/swerust1/pubs/ldlst5.pdf

Summary of Federal Regulations Governing USTs
http://www.epa.gov/swerust1/pubs/musts.pdf

Underground Storage Tanks: Requirements and Options
http://www.epa.gov/swerust1/pubs/ustsrao.pdf

Massachusetts Installation and Maintenance of USTs
http://www.state.ma.us/legis/laws/mgl/gl-148-toc.htm

ASBESTOS REGULATIONS

9

Charles Ciccone

What is asbestos? Asbestos is a mineral fiber that contains various components of silicon, oxygen, hydrogen, and various metal ions. It exists in many varieties, with most containing one of three components, chrysotile, amosite, and crocidolite. Together, these minerals are often referred to as asbestos-containing material (**ACM**). Chrysotile or "white asbestos," the most common, accounts for approximately 95% of asbestos found in buildings in the U.S. Asbestos derives from a Greek adjective meaning inextinguishable. It was initially thought a miracle mineral due to its ability to withstand heat and for its pliancy allowing it to be woven into cloth or wicks. The Greeks and Romans were purported to throw their asbestos tablecloths into fires for cleaning. For this reason, it has been widely used in the commercial, industrial, and construction industries with reports of over 3,000 different types of asbestos-containing products.

The commercial use of asbestos, chrysotile, dates back to the 1870s when the first mine opened in Quebec, Canada. Following the discovery of the large asbestos deposits in Canada, asbestos emerged as a preferred insulating material found throughout home and industry. Asbestos is commonly found in boilers, pipes, reinforcement materials, and brake-pad linings. A more complete list of materials can be viewed at the EPA Web site, *http://www.epa.gov/earth1r6/6pd/pd-t/asb.mat1.hem*. Its use is widespread, which has led to a number of different federal and state regulatory laws. The physical characteristics of asbestos include the formation of crystal and needle-like shapes that are easily inhaled and then deposited deep into the lungs. It is for this reason that despite the banning of asbestos and its gradual phase-out, it remains a significant threat to human health. Also, the health affects from asbestos may sometimes take many years to develop.

SUMMARY OF REGULATIONS/POTENTIAL HEALTH EFFECT

Many federal and state laws have evolved to control exposure, transport, disposal, and prevention of asbestos related illness. Epidemiological studies have demonstrated increased health risks associated with ACM. Due to the unique fiber structure of asbestos, its use has been associated with significant medical problems including pulmonary fibrosis (stiffening of the lungs), mesothelioma (a rare type of cancer of the lining of the lung or abdomen), pleural plaques, and lung cancer. The risk from lung cancer in a smoker of 20 cigarettes per day is approximately 10X and with asbestos exposure is 5X that of the general public. The risk of lung cancer from asbestos is exacerbated in smokers with both together magnifying the risk 50 fold. Additionally, many people who were exposed never knew that a possible exposure to asbestos related materials would include unique long-term medical effects. When the fibers from ACM are deposited deep into the lung, the body is unable to expel or destroy the fibers. This leads to a delayed effect due to the stiffening of the lungs, pulmonary fibrosis, where it becomes increasingly difficult to breathe. Mesothelioma and lung cancer are extremely difficult to detect and treat and are frequently fatal with a delay of as many as 20 to 30 or more years from initial exposure. The EPA believes there is no safe level of asbestos exposure. Due to the delay in disease onset, both asbestos workers and their families members who have been unintentionally exposed to ACM may still develop asbestos-related disease many years from now.

Both Federal and State laws address the health risks due to asbestos and ACM exposure. The main federal regulation for asbestos falls under the Toxic Substances Control Act (TSCA) of 1976 with its subsequent amendments and expansions. There are several other federal and state laws that restrict asbestos use, transport, storage, and disposal. The EPA regulations governing asbestos also fall under the Clean Air Act (CAA) of 1970 and the Safe Drinking Water Act of 1974. TSCA regulations and guidance are administered and managed by the Office of Pollution Prevention and Toxics (OPPT). Other important regulations include the Asbestos Hazard Emergency Response Act (AHERA) of 1986, the National Emissions Standards for Hazardous Air Pollutants (NESHAP), the asbestos ban and phase-out rule, and waste transport and disposal regulations. These will be discussed in greater detail separately.

Phasing Out ACM [1]

The Toxic Substance Control Act (**TSCA**) was enacted by Congress in 1976, to regulate many toxic chemicals and substances that posed an unreasonable risk of injury to health or the environment. Under TSCA, there are special provisions that regulate, restrict, and prohibit specific toxic substances. This includes asbestos, radon, PCBs, fully halogenated chlorofluorocarbons (**CFCs**), and dibenzodioxins and dibenzofurans. Other substances including pesticides, tobacco and tobacco products, radioactive materials, food, drugs, and cosmetics do not fall under TSCA but are instead regulated under the Federal Insecticide, Fungicide, and Rodenticide Act (**FIFRA**). TSCA promotes research into a database for toxic substances and regulates their safe transport via interstate commerce.

The EPA is authorized to prohibit the manufacturing, processing, importation, or distribution in commerce of toxic substances under Section 6 of TSCA. In 1979, the EPA first attempted to reduce the health risk from exposure to asbestos but final regulations were not issued until July 12, 1989. A document was published in 1989 by the EPA titled, *Asbestos: Manufacture, Importation, Processing, and Distribution in Commerce Prohibitions; Final Rule (40 CFR Part 763, Subpart I).* This rule will eventually ban about 94% of the asbestos used in the U.S. and be implemented in three stages between 1990 and 1997. The attempted ban and phase-out over an 8-year period did not occur as expected and in fact the U.S. Fifth Circuit Court of Appeals in 1991 set aside and vacated much of the phase-out. This misunderstanding of what is and is not banned is further clarified as of May 1999 under the EPA web site, *http://www.epa.gov*.

EPA asbestos regulations fall primarily under two federal laws, the Clean Air Act (NESHAP) and TSCA. To a lesser extent, the U.S. Consumer Product Safety Commission (CPSC) also has some bans to include textured paint and patching compounds. Thus, bans on some ACM products and uses remain as of April 1999. This includes under the CAA, most spray-applied surfacing ACM, wet-applied and pre-formed asbestos pipe insulation, and pre-formed asbestos. Under the TSCA, the asbestos ban precludes its use in corrugated paper, rollboard, commercial paper, specialty paper, flooring felt, and new uses. In summary, the EPA has no existing bans on most other asbestos-containing products or its uses.

Demolition and Removal Involving ACM [2]

All facilities which are to be demolished become subject to regulation under NESHAP. Demolition and renovation projects that include friable and non-friable asbestos involved in the intentional burning or wrecking of buildings are also specified in the 1990 NESHAP regulations. The EPA distinguishes friable and non-friable forms of ACM. Friable asbestos is any ACM that may be crumbled or reduced to powder by hand pressure. This is thought to more easily aerosolize asbestos fibers.

There are procedures that must be followed by the operator if the facility contains more than 80 linear meters of regulated ACM on pipes, 15 square meters on other areas, or a total of one cubic meter of regulated ACM. An owner or operator must:

1. Provide detailed information, including the name of the asbestos waste transporter and ultimately, the waste disposal site
2. Inspect the site for the presence of and estimate the amount of non-friable ACM
3. Describe the procedures to be followed in the event non-friable asbestos becomes friable in the course of the demolition or renovation
4. Notify DEP 10 days in advance, before the actual start of the work
5. Update the notice whenever there is a change in start date, which may be done via telephone but must include a written follow-up
6. Update the notice whenever the amount of regulated ACM changes by 20% from earlier predicted amounts
7. Provide the name, address, and telephone number of the facility owner, operator, and asbestos contractor; type of project
8. Provide the asbestos detection procedures used and estimated amounts of regulated ACM
9. Provide the scheduled projects' start dates and end dates
10. Provide a description of the facility and location
11. Provide the name and location of the waste disposal site
12. Provide descriptions of the work methods and engineering controls

Under provisions of Asbestos NESHAP, a trained person must supervise operations where ACM is handled, stripped, or removed. They are ultimately responsible for any on-site activity and are there to ensure proper compliance with the laws and prevention of friable ACM originating from non-friable asbestos. Handling of ACM is regulated to keep air emissions to a minimum. All regulated ACM must be removed in a project containment area with the exception of either encased or non-friable materials. Cautious handling and wetting of waste ACM helps to ensure prevention of airborne particulates. During freezing temperatures, measures must be undertaken to ensure asbestos removal is suspended when wetting of asbestos is being performed and the asbestos contractor must record three daily temperature readings scattered throughout the workday. Other particulate control measures include the use of local exhaust vents, glove bag systems or leak-proof wrapping when wetting is not required. Records must then be maintained for at least two years.[3]

Asbestos standards for the construction industry exist mostly under the Code of Federal Regulation 29 (CFR). This includes under general industry, 29 CFR 1910.1001, the construction industry under 29 CFR 1926.1101, and finally, the use of respirators under 29 CFR 1910.134.

EPA Worker Protection Rule

The EPA's Worker Protection Rule (40 CFR Part 763, Subpart G) parallels and extends OSHA standards to state and local employers performing asbestos work who may not be otherwise covered. This includes requirements similar to the Occupational Safety and Health Administration (**OSHA**), which requires and covers the cost for medical and air monitoring, protective equipment, work practices, and record keeping. States and local agencies may also have more stringent requirements than those required by the federal government. Both the EPA and OSHA work in concert with each other. The EPA is responsible for regulating environmental exposure while OSHA is responsible for the health and safety of workers who may be exposed to ACM in the workplace.

Regulation of asbestos inspectors is important in the monitoring of an asbestos program. They may face litigation in three areas of liability: contractual liability, tort negligence, and regulatory-type. In contractual liability, inspectors are liable for any breach of contract if services are not properly performed. Tort litigation deals with negligence, either intentional or unintentional. Finally, regulatory liability involves non-compliance with either state or federal regulations.

Transport and Disposal of Asbestos Waste [2]

Additional restrictions exist concerning ACM (asbestos containing material) disposal, some of which fall under the 1990 NESHAP regulations. There can be no discharge or visible emissions to the outside air in the manufacture, fabrication, demolition, renovation, or spraying of ACM. Waste of asbestos material must be clearly marked, wetted, and placed into leak-proof containers, which must have an OSHA-specified label along with the name of the waste generator.

The vehicles used to transport ACM must be marked according to guidelines during both loading and unloading. Labels with the name of the waste generator and its location must be placed on all containers. The generator is required to maintain waste shipment records (WSR) similar to hazardous waste manifests. A copy of the waste shipment record is sent to the disposal site owner or operator along with the ACM waste. The owner or operator then must send a copy of the WSR back to the generator within 30 days. At this time, any discrepancies in the quantity of waste are reconciled from what's on the WSR and the amount received. If there is a discrepancy, it must be resolved within 15 days of receiving the waste or written notification occurs to the agency. This is usually a NESHAP administering agency such as a state environmental agency, which in Massachusetts is the Mass. Department of Environmental Protection (**DEP**).

New and existing ACM disposal sites are strictly monitored and approved. They must conform to existing emission standards and controls and provide the EPA with the name and address of the operator, the location of the site, and the average weight per month of hazardous materials processed. Additionally, a copy of the WSR signed by the waste site must be returned to the waste generator within 35 days of the date that the waste was accepted. If not then the initial ACM generator must contact the transporter and/or waste site owner/operator to determine the status of the shipment. If the signed copy is still not received within 45 days, then the waste generator must submit a written report to the administering agency.

Waste disposal sites owners need clear records on deeds to the property indicating that the site has been used for ACM disposal. The records must be detailed to include location, depth, and volume of asbestos waste. Inactive site owners will need written approval before excavation or disturbance of any site used for ACM disposal.[4]

ASBESTOS IN PUBLIC BUILDINGS AND SCHOOLS- Evolution of the Asbestos Hazard Emergency Response Act (AHERA) [2]

The EPA has an ongoing effort to improve methods and offer guidance for controlling exposure to asbestos in buildings. In a study by EPA in 1988, asbestos was found in nearly one fifth of all public and commercial buildings. This number was so great that any comprehensive removal program would overwhelm the existing asbestos abatement system. Therefore, the EPA has recommended the following:

1. Broader training in asbestos control
2. Focusing of attention on the most dangerous thermal system insulation asbestos
3. Better coordination of the proliferating asbestos abatement activities
4. An evaluation of the effectiveness of AHERA.

In 1986, the Asbestos Hazard Emergency Response Act (**AHERA**; Asbestos Containing Materials in Schools, 40 CFR Part 763, Subpart E) was signed into law as Title II of TSCA.[4] It supercedes the previous Asbestos-in Schools Rule promulgated in 1982. AHERA was essentially developed to deal with the extensive problem of asbestos in schools and other public buildings in which nearly one fifth were found to contain some form of asbestos materials. Local educational agencies (**LEAs**) under AHERA inspect their schools for ACBM and then prepare the best possible plan to effectively deal with the problem. This may include repairing damaged ACBM, sealing, enclosing, or removing it or trying to keep it in good enough condition to prevent the formation of airborne fibers. Approval of the plans by accredited management planners occurs with oversight by the state. The LEAs have the responsibility to notify employers, parents, and teachers of any plans. These plans are then implemented through AHERA and accredited abatement designers, contractor supervisors, workers, building inspectors, and school management plan writers. The regulatory agencies responsible for AHERA concentrate their efforts through educating the local education agencies in order to ensure compliance. Asbestos improperly removed from a school may result in liability to the contractor through both AHERA and NESHAP regulations. All public, private, and secondary schools were to be inspected for ACBM by June 28, 1983.

NATIONAL EMISSION STANDARDS FOR HAZARDOUS AIR POLLUTANTS (NESHAP) [4]

The Clean Air Act (CAA) of 1970 contains EPA rules on the application, removal, and disposal of asbestos. This is to protect the public form the airborne risk of asbestos contaminants that are known to be hazardous to human health. Under Section 112 of the CAA, EPA established the National Emission Standards for Hazardous Air Pollutants (NESHAP), with asbestos being one of the first recognized hazards on March 31, 1971. EPA promulgated the Asbestos NESHAP in 40 CFR Part 61, Subpart M. The most recent amendment after several iterations is form November 1990.

Many different sources are regulated under NESHAP. They include:

1. The milling of asbestos
2. Roadways containing ACM
3. The commercial manufacture of products that contain asbestos
4. The demolition of all facilities
5. The renovation of friable asbestos in facilities
6. The spraying, processing, and use of asbestos-insulating materials that contain ACM
7. The disposal of asbestos-containing waste
8. The monitoring of closed or inactive asbestos sites
9. Facilities which convert asbestos into non-asbestos material
10. Air-cleaning devices in their design and operation
11. The reporting of information on process and filter equipment
12. Active asbestos waste disposal sites

Ultimately, the goal of NESHAP is to prevent the release of asbestos fibers during any activity which may involve asbestos. The regulations require any owner to notify the state or local agencies and/or the EPA before any demolition or renovation can take place. Not only does NESHAP regulate asbestos through waste handling and disposal, it also restricts the use of any type of spray asbestos or use of the wet-applied or molded insulation such as that found on pipes.

STATE AND LOCAL REGULATION OF ASBESTOS

Many individual states have two or more agencies which regulate asbestos. Frequently the state often oversees the training, licensing, and certification of contractors, supervisors, and workers often falling under the State Department of Labor. Asbestos disposal issues are frequently regulated by the department, which controls environmental protection or conservation. The following are examples of a few specific state programs.

Massachusetts [2]

Two agencies are responsible for an asbestos program in Massachusetts. The Department of Labor and Industry (DLI) licenses contractors and workers and regulates workplace standards. This program provides regulation of all work, construction, demolition, alteration, repair or maintenance at any facility using, handling, or disposing of asbestos. DLI certifies asbestos training courses, licenses contractors and their employees, and develops and enforces workplace standards. The Department of Environmental Protection (DEP) monitors asbestos disposal and inspects workplaces.

New York [5]

In New York, asbestos is regulated by two sets of laws. The New York State Labor Law (the Labor Law) regulates contractors who handle and remove asbestos. It also specifies education and training for asbestos workers. There is a mandatory requirement for training in the handling of asbestos for any employee who comes into contact with ACM. Licensing and certification is provided under these regulations for contractors, supervisors, and employees successfully completing various levels of training. These requirements vary somewhat among states. The New York State Department of Labor (DOL) enforces the Labor Law. The second set of asbestos related laws in New York is through the N.Y. State Education Law, which is administered by the Commissioner of Education. It requires inspection of the New York public schools every three years and is patterned after the federal Asbestos Hazard Emergency Response Act (AHERA).

New Jersey [6]

There are four organizations responsible for asbestos regulations in New Jersey. The New Jersey Department of Health and the Department of Labor have the authority to train and license both asbestos workers, who are required to have a permit, and contractors, who are required to have a license. The New Jersey Department of Community Affairs enforces sections of the Uniform Construction Code dealing with asbestos work, while the New Jersey Department of Environmental Protection and Energy has responsibility for regulating the disposal of asbestos wastes.

REFERENCES

1. **Pub. L. No. 100-694**, 102 Stat. 4563 (Also referred to as the "Westfall Act"), codified at 28 U.S.C. 2679, 1988.

2. **Vance, B., Weinsoff, D.J., Henderson, M.A., and Elliot, J.F.**, *Toxic Program Commentary: Massachusetts*, Specialty Technical Publishers, Inc., North Vancouver, B.C. Canada, 1992.

3. **United States Environmental Protection Agency,** *"The Asbestos Informer,"* Region 4, Air, Pesticides, and Toxics, June 1999. http://www.epa/region4/air/asbestos/inform.htm

4. **"The Asbestos Informer,"** U.S. Environmental Protection Agency, Region 4, Air, Pesticides, and Toxics, June 1999. http://www.epa/region4/air/asbestos/inform.htm

5. **Vance, B., Weinsoff, D.J., Henderson, M.A., and Elliot, J.F.**, *Toxic Program Commentary: New York,* Specialty Technical Publishers, Inc., North Vancouver, B.C. Canada, 1993.

6. **Weinsoff, D.J., Demes, J., Polkabla, M.A., Henderson, M.A., and Elliot, J.F.**, *Toxic Program Commentary: New Jersey*, Specialty Technical Publishers, Inc., North Vancouver. B.C. Canada, 1993.

WEB LINKS FOR THIS CHAPTER

Asbestos FAQ
http://www.epa.gov/region04/air/asbestos/inform.htm

Asbestos Home Page
http://www.epa.gov/opptintr/asbestos/index.htm

Asbestos Laws and Regulations
http://www.epa.gov/opptintr/asbestos/asbreg.htm

Sample List of Suspect ACMs
http://www.epa.gov/earth1r6/6pd/pd-t/asbmatl.htm

Massachusetts Asbestos Abatement
http://www.magnet.state.ma.us/dep/matrix.htm

Reporting and Recordkeeping Requirements for Asbestos Waste
Materials
http://www.epa.gov/region04/air/asbestos/waste.htm

PESTICIDE REGULATIONS

10

Sherri McGloin, Ian Cambridge, and Evan Hutchinson

Prior to 1970 when President Nixon created the Environmental Protection Agency (EPA) through executive order, the government's involvement in the protection of human health and the environment from pesticides was minimal. The federal government focused on the targeting of manufacturers to reduce the production of ineffective pesticides and deceptive mislabeling. In 1910 the government created the Insecticide Act. The act mandated that the manufacturer register, produce, and label effective pesticides.

Enforcement of pesticide protection was the responsibility of the U.S. Department of Agriculture (USDA; <*http://www.usda.gov*>) and in 1947 the department established the Federal Insecticide, Fungicide, and Rodenticide Act (**FIFRA**). The USDA required registration of pesticides and set a standard for labeling requirements. The USDA did not establish guidelines for pesticide use. Therefore all pesticides that were registered were used regardless of their potential to be hazardous to humans or the environment. The manufacturer was not liable for negative impacts that the pesticide might produce.

It was the environmental movement of the 1960s that prompted the government to react in a more responsible way to the regulation of pesticides. The public became aware of the effects of agricultural pesticides and brought their concerns to the government. The public demanded suspension and immediate cancellation of pesticides that had shown severe damaging effects to humans and the environment such as DDT, Aldrin-Dieldrin, Mirex, and the herbicide 2,4,5-T.

In 1970 Earth Day was celebrated and President Nixon signed the Reorganization Order No. 3, giving rise to the Environmental Protection Agency (EPA). The EPA inherited the FIFRA responsibility from the USDA. In conjunction with the concerns of pesticide regulations, the EPA interacts with the Food and Drug Administration (FDA) and the Occupational Safety and Health Administration (OSHA).

The impact of the EPA prompted the National Environmental Policy Act (NEPA) to be signed into law in order to provide analysis of the environmental impacts of federal actions. The president's Council on Environmental Quality (CEQ) was established to advise and assist the President on environmental policies.

By 1972 FIFRA was not adequately monitoring the health and environment and was amended by the Federal Environmental Pesticide Control Act (FEPCA) to reevaluate the laws under FIFRA with a three-year deadline. FIFRA was amended in 1975. The 1975 amendment didn't add any new changes to the written act, but it did give direction toward where the FIFRA was headed. Congress decided to restrict the EPA's authority by eliminating the power of final approval on passing pesticide laws.

The 1978 and 1980 amendments implemented the cause for substantial studies by the manufacturers through the Scientific Advisory Committee to research the potential outcome a pesticide might have on humans and the environment. The committee had the power to reject or cancel any pesticide under their provision.

One area of concern that hadn't been addressed by FIFRA was the contamination of ground water by pesticide applications. President Ronald Reagan in 1988 signed the 5th amendment into FIFRA to protect and prevent the contamination of ground water supply. Two more amendments followed again in 1990 and in 1996.

On August 3, 1996 President Clinton signed a rewritten version of a federal pesticide regulation that involved measures that did away with the 1958 Delaney Clause.[1] The Delaney Clause of the 1958 Federal Food, Drug, and Cosmetic Act (*<http://www.fdc.gov>*), which prohibited processed foods with any trace of any chemical that was proven to cause cancer. The new provision says the EPA must ensure that pesticide residues on both processed and raw foods do not pose more than a one in a million chance of causing cancer.[7] This amendment in addition to setting more stringent standards for foods, prompted the EPA to instruct the Department of Agriculture (**USDA**) and the Health and Human services to study children's eating habits and pesticide exposure.[2] Pesticides found on fruits and vegetables, including Organochlorine (OP), Pyrethroid, and herbicide compounds, are considered to have potential adverse health effects on children.[3]

Pesticide Registration

Pesticides are defined by the EPA (*http://www.epa.gov/pesticide*), as "any substance intended for preventing, destroying, repelling or mitigating any pest, any substance or mixture of substances intended for use as a plant regulator, defoliant, or desiccant."[4] To submit a pesticide for registration to the EPA the manufacturer must supply significant studies reflecting the safety of the pesticide. In addition the appointed administrator of the agency can ask for additional information. In some cases this process can take years. At the federal level a fee is assessed for each active ingredient. The guidelines for pesticide registration are becoming stricter due to the enforced protection of human consumption and land protection.

Suspension and Cancellation of Pesticides

Suspension of a pesticide calls for an immediate ban on the production or distribution (*<www4.law.cornell.edu/uscode/7/136d.text.html>*). Ordinary suspension is involved with the cancellation of a pesticide that is being reviewed. Emergency suspension is the strongest action by the EPA that can be taken under FIFRA. It calls for immediate stops of all use, sale, and distribution of pesticides. Once a suspension is in place the registrant shall have an opportunity to have an expedited hearing before the administrator on whether an imminent hazard exists. A request for a hearing must be submitted within five days of notification of suspension or the order will take place immediately and will not be reviewable by a court.[6]

A pesticide that is suspected of posing a question of safety to human beings and/or the environment is subject to scientific testing, review, and possible cancellation.

Pesticide Storage and Labeling

All pesticides that hold a potential adverse effect on the environment and humans should meet storage facility regulations:

1. Designated site selection
2. Protective enclosures
3. Highly toxic or moderate toxic should be labeled bearing the words: DANGER, POISON, or a warning that is plainly visible
4. Flooding is unlikely
5. Soil texture/structure and geologic/hydrologic characteristics will prevent the contamination of any waste by runoff or percolation
6. Secure climb-proof fencing
7. Identification signs should be visible warning of their hazardous nature
8. All equipment used in handling pesticides should be labeled "contaminated with pesticides"

Containers holding pesticides must be checked regularly for leaks and corrosion. A list of accident prevention measures must be provided at each storage site along with safety measures. A monitoring system should be considered in the location of the storage facilities. Samples of surface and ground water, wildlife, and plant environment should be tested regularly. Analyses should be performed according to "Official Methods of the Association of Official Analytical Chemists (AOAC)." [5]

Disposal of Pesticides

The disposal of pesticides is based on the structure of the compound. Organic and metallo-organic have similar procedures for disposal, except that metallo-organic has to be chemically and physically treated to remove heavy metals from the hydrocarbon structure. Incineration is recommended for disposal of pesticides if possible, but must be in compliance with the Air Quality Act established in 1970. Landfills are designated for pesticides under certain guidelines. All disposals of pesticides must be approved by the agency administrator along with the proper disposal of toxic pesticides under the Toxic Substance Control Act (TSCA).

Disposal containers are classified as:

1. Group I Containers: Combustible container
2. Group II Containers: Non-combustible container

3.   Residue Disposal: Residue and rinse liquids

Penalties

Pesticide penalties fall under U.S. code: Title 7, Section 136l, stating any distributor who violates any provisions under the subchapter of Title 7 will be subjected to civil penalties under the administrator of not more than $5000 for each offense. Any private person can receive up to a $1000 for each offense. All cases are subjected to a hearing in the court of law. Many violations of pesticide regulations fall under civil penalty. Each state can and does set its own penalties subjected also to a court of law, but in many cases the fines can be up to $25,000.

Criminal penalties fall under similar regulations with higher fines for the registrant. Title 7 lists fines to be not more than $50,000 or imprisonment for not more than one year.

The state of Massachusetts clearly states that any person who violates pesticide regulations shall be fined $25,000 or imprisonment for not more than one year under a civil penalty, noting that each day of violation shall constitute a separate offense. [6]

REFERENCES

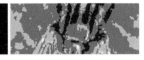

1. **Miller, M. L.**, *Environmental Law Handbook*, Chapter 12, Pesticides, 15th ed., Government Institutes: a Division of ABS Group Inc., 1999.

2. **Bradman, A., Castorina, R., and Eskenazi, B.**, *Exposures of Children to Organophosphate Pesticides and Their Potential Adverse Health Effects*, Environmental Health Perspectives, Vol. 107/3/supp. June, 1999.

3. **EPA**, *Endocrine Disruptor Screening Program: Proposed Statement of Policy*, December 28, 1998.

4. **EPA**, *Pesticide Programs, http://www.epa.gov/pestpg*, Spring 2000.

5. **EPA**, *Pesticide Regulations and Laws, http://www.epa.gov/pestpg*, March 2000.

6. **U.S. Code**, Cornell Law School, *http://www4.law.cornell.edu/uscode/7/1361*, Spring 2000.

7. **PBS**, "Online News Hour: Environment Backgrounder – Re-Regulation of Pesticides," *http://www/pbs.org/newshour/backgrounders/pesticides.html*, Fall 1996.

WEB LINKS FOR THIS CHAPTER

EPA Office of Pesticides Programs
http://www.epa.gov/pesticides

FIFRA Summary
http://www.epa.gov/region5/defs/html

Full Text of FIFRA
http://www4.law.cornell.edu/uscode/7/ch6.html

Information Resources and Links
http://www.epa.gov/pesticides/info.htm

Massachusetts Pesticide Control Act
http://www.state.ma.us/legis/laws/mgl/gl-132B-toc.htm

Massachusetts Pesticide Bureau Index Page and Links
http://www.massdfa.org/pesticideindex.htm

Toxics Pesticides Enforcement Division
http://www.epa.gov/envirosense/oeca/ore/tped

AIR QUALITY REGULATIONS

11

Gary Moore and Michael Pepe

T he problems associated with air pollution are not recent. The obnoxious stench of burning coal was so annoying to Edward I and Edward II of Great Britain, that those caught burning coal were at risk of losing their lives. In fact, a man was hanged during the reign of Edward II for burning coal and fouling the air.[1] Such rigorous penalties haven't been witnessed since then. However, air pollution has continued to be a problem that worsened with the combined effects of an increasing population and the industrial revolution.

Generally, air pollution was considered to be an annoyance, resulting in respiratory irritation, deposits of dust and filth on surfaces, and reduced visibility. However, the potentially severe consequences of air pollution were thrust into the vision of the public eye in documented episodes that have become "classics" in the air pollution literature. These episodes began in the first week of December in 1930 in the Meuse River Valley, Belgium. This deep river valley was highly industrialized with manufacturing plants in steel, glass, lime, fertilizers, and sulfuric acid. That week in December, the air became stilled, and the continuously emitted pollutants in the valley were trapped. The residents complained of irritated throats, coughing, and tightness in the chest. More than 60 deaths were attributed to the episode, primarily among the elderly and those with pre-existing cardio-respiratory disorders.

Nearly 18 years later in Donora, Pennsylvania, similar conditions prevailed with steel manufacturing concentrated in a river valley. A stable, slow-moving air mass resulted in the formation of a dense, cool layer of air near the earth with warmer air above in a phenomenon known as an inversion. The noxious pollutants of industry were trapped below the interface of the warm air above and the cool air below, and began accumulating on Wednesday, October 28th. The first death from this episode occurred on that Saturday and nearly 20 people died by Sunday evening. Six thousand people, or 40% of the population, developed symptoms associated with the episode including nausea, vomiting, irritation to the eyes, nose, and throat, headaches, and muscular aches and pains. The severest symptoms and the most deaths were among the elderly and those with pre-existing respiratory and cardiovascular conditions. The lungs of those who died showed evidence of capillary dilation, hemorrhaging, purulent bronchitis, and edema.[2,3]

Excess deaths were also demonstrated for London, England during several air pollution episodes associated with stagnant air. These episodes occurred in 1952, 1956, 1957, and 1962. London's worst episode was in 1952 during which a thick yellow smog settled over the city for five days, resulting in the deaths of 4,000 more people than the number expected in the absence of air pollution. The deaths were mostly among the elderly and those with pre-existing respiratory or cardiac problems. The excess deaths spiked with the air pollution episode and subsided when the the stagnant air was eventually swept away by more turbulent conditions.

Episodes of air pollution resulting in measurable levels of excess deaths have also occurred in Los Angeles and New York, primarily in the 1950s and 1960s. The air pollution disasters focused public attention on the severity of air pollution. It became clear that exposures to concentrated levels of ambient pollutants could cause severe symptoms and even death. Such awareness prompted greater attention to studying the factors associated with air pollution and then establishing regulations to control and limit the pollutants. Such efforts have made it unlikely that similar air pollution episodes will occur in industrialized countries. However, countries such as China, Mexico, and eastern Europe are facing substantial air pollution problems as they strive for industrial growth in the absence of vigorous air pollution control policies.

Although progress in controlling air pollution in the developed nations has been made, problems of global concern remain. Government leaders and scientists are building toward a consensus in the midst of controversy that industrial emissions such as carbon dioxide, and other greenhouse gases, are contributing to global warming with potentially severe consequences. Emissions of nitrogen and sulfur oxides are thought to be major contributors to acid deposition, which is being associated with the disappearance of fish and other biota in the tens of thousands of lakes and streams worldwide, and the decimation of high altitude forested areas. Chlorofluorocarbons, once widely used in air conditioners, in electrical solvents, and in the production of foam cups, have been linked to to the catalytic destruction of stratospheric ozone in a process known as ozone depletion. Such depletion is thought to result in higher levels of damaging ultraviolet radiation reaching the earth. These atmospheric effects of air pollution have mobilized governments internationally to meet and attempt to resolve these issues. Actions that may be taken are likely to have far-reaching effects on economies, standards of living, and the livelihood of hundreds of millions of people. The quality of our lives, the stability of our environment, and the survivability of the planet have been brought forward as arguments in these controversial issues. Is there a global warming problem? Is stratospheric ozone being depleted? Is acid deposition killing forests internationally and destroying aquatic life? These issues must be weighed against the increased regulatory demands for clean air.

Air pollution not only represents a perceived threat to global ecological damage, but also exerts both acute and chronic threats to human health at exposures within the normal range of pollutants in our cities and towns.[4] These threats include respiratory disorders such as asthma, bronchitis, and emphysema. There are also increased risks of cardiovascular disease, cancer, and respiratory infections. Air pollutants have been implicated in direct injury to living plants, and to materials including: (1) the corrosion of metal, (2) soiling of buildings, and (3) the degradation of paints, leather, paper, textiles, and dyes. These effects result in significant economic losses approaching several billion dollars annually.[5] Based on the overview given above, the arguments for controlling air pollution are substantial.

THE CRITERIA POLLUTANTS
Introduction

The USEPA has reported substantial progress in improving the air quality in many states, including the heavily polluted areas of Southern California, Hawaii, and regions in Arizona and Nevada. Despite major increases in population and acute travel over the past decade, air pollution has decreased by about one-third overall in these regions. The greatest reductions have been recorded for lead (93%), followed by carbon monoxide (35%) and particulate matter (26%). While the majority of pollutants are being steadily reduced, nitrogen oxides, ground-level ozone and fine particulates remain a problem in many areas.[6] A listing of the criteria pollutants and current standards is provided in Table 11-1.

The total emissions of criteria pollutants fell 32% since 1970 in the face of a 29% increase in the U.S. population, a 121% escalation in vehicle miles traveled, and a gross domestic product increase of 104%.[6] Still, nearly 46 million people live in counties that fail to meet the air quality standards for one or more of the criteria pollutants.[6] Fine particulates from power plants, motor vehicles, and photochemical reactions in the atmosphere kill as many as 64,000 Americans each year from lung and heart disease.[7] Similarly, the American Lung Association released a survey of 13 U.S. cities linking exposure to ground-level ozone (smog) to as many 50,000 emergency room visits and 15,000 hospital admissions.[8] The health and welfare problems associated with criteria pollutants have not been completely resolved and many issues remain. A summary of criteria pollutants sources and health and welfare effects is presented in Table 11-2.

Particulate Matter (PM)

Particulate pollutants include airborne particles in liquid-solid form that range in size from visible fly ash greater than 100 μm to particles 0.005 μm in size (Table 11-3). Particles may be produced naturally, such as pollen or sea spray, or by human activities, such as industrial processes, agricultural activities, fossil fuel combustion, and traffic. Particulates include dust, smoke, soot (carbon), sulfates, nitrates, trace metals, and condensed organic compounds. Particulates produce a number of effects adverse to human interests including: (1) respiratory and cardiac health hazards to humans; (2) the deposit of grime and soot on buildings; (3) the reduction in sunlight, thereby causing a regional or global coding effect; and (4) reduced visibility in areas of extensive smoke or particulate pollution.

The characteristics for coarse particles (2.5 to 10 μm) are very different from fine particles (<2.5 μm) and relate to the adverse effects they produce. Coarse particles originate from wind-blown dust coming from deserts, unpaved roads over which vehicles

travel, agricultural fields, pollen, mold spores, and plant debris. These particles tend to be mostly minerals consisting of aluminum, silicon, potassium, and calcium, which are chemically basic. Fine particles (<2.5 μm) are mostly emitted from fossil fuel combustion in industry, residences, and motor vehicles. They are also created in the atmosphere from gases of nitrogen and sulfur oxides and volatile organic
compounds forming sulfates, nitrates, aliphatic and aromatic hydrocarbons including aldehydes, ketones, phenols, esters, terpenes, and phenylacetic acids.[9,10] Coarse particles can build up in the respiratory system causing respiratory disorders such as asthma in women. Smaller particles are more apt to be associated with: (1) increasing lung

POLLUTANT	AVERAGING TIME	PRIMARY STANDARD	MAIN SOURCES
Carbon monoxide	8 hours	9.0 ppm	Transportation
Hydrocarbons (corrected for methane)	3 hours (6-9 am)	160 μg/m^3	Transportaion, Industrial processes
Nitrogen dioxide	Annual average	0.05 ppm	Stationary source fossil fuel combustion, Transportation
Sulfur dioxide	Annual average	0.03 ppm	Stationary source fossil fuel combustion
Particulates PM$_{10}$	Annual arithmetic mean	50 μg/m^3	Multiple sources including stationary source fuel combustion, industrial processes, and transportation.Fine particles are associated with photochemicals and products of fossil fuel combustion
PM$_{2.5}$*	Annual arithmetic mean	15μg/m^3	
	24 hr. average	65μg/m^3	
Ozone	1 hour 8 hours•	0.12 ppm 0.08 ppm*	Secondary pollutant formed in presence of sunlight, NOx, and hydrocarbons. Fossil fuel combustion is a major contributor
Lead	3 months	1.5 μg/m^3	Food, dust, older houses with lead paint

•July 1997 amendments to the 1990 CAAA, being contested in court.

Table 11-1. National ambient air quality standards - criteria pollutants

CRITERIA POLLUTANTS AND SOURCES	HEALTH EFFECTS	WELFARE EFFECTS
Carbon monoxide Incomplete combustion of fossil fuels as in vehicles, kerosene heaters, boilers, and furnaces. Cigarette smoking, forest fires, and biological decomposition.	Interferes with oxygen transport in blood by binding with hemoglobin, Causes headaches, fatigue, cardiovascular disease, and central nervous system disorders.	Effects on plants or materials are not evident.
Nitrogen dioxide Emitted from the combustion of fossil fuels in vehicles, industrial boilers, and electric generating utilities.	Causes increased risk of respiratory infections and aggravates symptoms in persons with asthma and chronic bronchitis.	Produces a reddish brown haze over cities, which reduces horizon visibility, causes leaves to yellow, and is a precursor to acid deposition and tropospheric ozone.
Sulfur dioxide Fossil fuel combustion especially in coal-burning electric power utilities, metal smelters, oil refineries, and industrial boilers.	Causes irritation of the throat and lungs, and aggravates symptoms in persons with asthma and chronic bronchitis.	Causes corrosion and deterioration of metals, brittleness of paper, paint discoloration, damages textiles and leaves of plants, and is a precursor to acid depostition.
Particulates PM_{10} and $PM_{2.5}$ Fossil fuel combustion emissions, industrial processes, photochemical reactions in atmosphere, mechanical abrasion.	Aggravates asthma, heart disease, and chronic lung disease. Alters lungs natural cleansing mechanisms. Smaller particles associated with the most severe symptoms.	Causes soiling of materials. grime deposits, and reduced visibility. Particulates from volcanic eruptions can reduce solar energy and produce temporary cooling effect.
Ozone A product of NOx emissions from motor vehicles, power utilities, and industries burning fossil fuels, combined with hydrocarbons and sunlight in the atmosphere.	Causes breathing difficulty, irritation to mucous membranes, and increases risk to respiratory infections. Acute exposures cause respiratory pain, bronchoconstriction, lung edema, and abnormal lung development.	Corrodes rubber, paint, weakens fabrics, rubber, and produces leaf damage and retardation of plant growth.
Lead Historically emitted from vehicles burning leaded gasoline. Emissions have been reduced by 98% since 1974. Most lead exposures in U.S. today are not airborne.	Damage to nervous system, blood forming tissues, kidneys. Evidence of neurobehavioral disorders including learning disabilities, and antisocial behavior.	No known effect on vegetation or materials.

Table 11-2. Criteria pollutants: sources and health effects

function in persons who have pre-existing conditions such as asthma; (2) increasing premature death among the elderly and those with cardiopulmonary disease; (3) deterioration of respiratory defense mechanisms and adverse changes in lung tissue and structure; and (4) increased respiratory disease and symptoms.[9]

The recent community studies provided support for issuing new standards for particulates in order to protect the public health and welfare. The new standard of PM 2.5 is set at 15 $\mu g/m^3$ annual arithmetic mean, and 6.5 $\mu g/m^3$ for a 24-hour average since these fine particles are more closely linked to mortality and morbidity effects than the previous PM10 standard.

Ozone and the Photochemical Oxidants

There are two types of ozone, which are known on the popular media as "good" ozone and "bad" ozone. Good ozone is that layer of ozone in the stratosphere previously discussed which absorbs UV-B radiation and protects the earth from excess amounts of these damaging rays. The 'bad' ozone is formed on the troposphere (nose-level) by a complex series of reactions including sunlight, nitrogen dioxide, and volatile hydrocarbons. The reaction follows that outlined in Fig. 11-1.

The photochemical oxidants resulting from such reactions include primarily ozone, nitrogen oxides, and alkyl peroxy radicals (RO2). The alkyl peroxy radicals are produced by oxidation of hydrocarbons. An oxidant is a substance that readily gives up an oxygen atom, or removes hydrogen from a compound. Photochemical refers to the ini-

PARTICULATE SIZE	PARTICULATE SOURCES	CHEMISTRY
Coarse Particles These are coarse particles from 1-100 μm including the course fraction PM_{10} (2.5-10 μm) .	Industrial and mechanical processes such as fragmentation of matter and atomization of liquids.Agricultural and forestry activities, and dust from unpaved roadways, mold spores, and wood ash.	Silicon, aluminum, iron, potassium, and calcium are common components. The coarse particle samples tend to be alkaline.
Fine Particles These are fine particles less than 1 μm, but including much of the mass of the $PM_{2.5}$ fraction.The $PM_{2.5}$ fraction consists of fine particles less than 1.0 μm and some coarse particles in the 1-2.5 μm range.	Industrial and residential combustion of fossil fuels. Secondary particles produced by direct, catalytic, and photochemical oxidation of nitrogen and sulfur compounds, and volatile hydrocarbons to produce sulfates, nitrates, and oxyhydrocarbons.	Elemental and organic carbon such as from fuel combustion (soot), sulfates, nitrates, condensed organic compounds, oxyhydrocarbons, and trace metals. The fine particle samples tend to be acidic.

Table 11-3. Sources and characteristics of coarse and fine particles

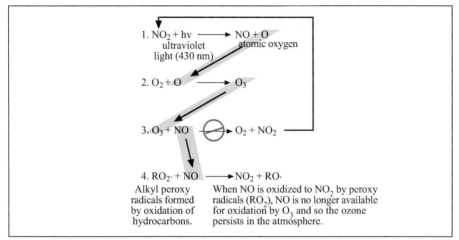

Figure 11-1. Proposed mechanism for the production and maintenance of tropospheric ozone (Adapted from Godish [4,2])

tiation of these reactions by sunlight. This mixture of photochemical oxidants is often referred to as photochemical smog and is most concentrated in areas with heavy traffic, intense sunlight, and stable air conditions. Such conditions are especially prominent in Southern California and to a lesser degree along the northeast coast, especially during May through September. Ozone is the primary indicator of smog and in concentrated form is a dense (1.6 times as heavy as air), violet blue gas with a sharp metallic odor. It is a powerful oxidant capable of breaking molecular bonds and rapidly degrading many structured materials, and plant and human tissues.

The ozone standard was revised in 1979 to be 0.12 ppm for one hour. However, more than 3,000 studies since the 1980s show evidence that adverse health effects occur at lower exposures than the previous standard, and that these symptoms are aggravated by longer exposures. Exposures to ozone are associated with increased hospital admissions from asthma, and also account for up 20% of summer time admissions of respiratory distress. Respiratory infections and inflammation are aggravated by ozone as well. Children and adults who play and work outside during summer months are at elevated risk to ozone exposure and may experience coughing, chest pain, and a reduction in lung function.

Exposure to ozone over long periods has been shown to cause inflammation of the lungs, impairment of lung defense mechanisms, premature aging of lung tissue through irreversible damage, chronic bronchitis, and emphysema. Consequently, the USEPA has attempted to set a new eight hour standard at 0.08 ppm based on the decision that this level will protect the public health,[11] but recent court challenges have placed the enforcement of those standards in jeopardy.[12]

Carbon Monoxide

Carbon monoxide is a colorless, odorless, tasteless gas that is produced from the incomplete combustion of fossil fuels with the greatest contribution by far being motor vehicles. In 1993, motor vehicles accounted for nearly 75% of anthropogenic carbon monoxide emissions.[10.] Between 1986 to 1996 carbon monoxide concentrations in ambient air have decreased 37 percent while overall emissions decreased by nearly 20%. These improvements were made while vehicle miles traveled increased 28% during the same period.[13]

Carbon monoxide enters the bloodstream through the lungs and combines with hemoglobin of red blood cells to form carboxyhemoglobin. Carbon monoxide has a stronger affinity for hemoglobin than oxygen, therefore as levels of carboxyhemoglobin rise, the adverse effects associated with oxygen deficiency increase. The health threat from chronic low level exposure is most serious for people who have pre-existing cardiovascular disease.[14] Inhalation of slightly elevated levels of carbon monoxide can produce symptoms ranging from dizziness and headaches to visual impairment, reduced work capacity, poor learning ability, lowered manual dexterity, and difficulty in performing complex tasks. Higher levels of CO inhalation can lead to paralysis of motor function and death.[13]

Lead

Lead is a systemic heavy metal poison which enters the human body through inhalation of contaminated air, and ingestion of lead in food, water, soil, or dust. Lead accumulates in the bones, blood, and soft tissues with target organs being the central nervous system, the blood forming tissues, and the gastrointestinal tract. Lead exposures have been associated with neurological impairments, mental retardation, behavioral disorders. Even at very low doses, children and fetuses suffer from central nervous system damage. The association of lead with behavioral problems and reduced intellectual ability caused lead to be placed on the list of criteria pollutants in 1977 when the Clean Air Act was re-authorized. Since automobiles burning leaded gasoline were the largest contributor to lead exposure, the phase-out of leaded gasoline has been the predominant control strategy. Lead emissions from highways have decreased 99% since 1987 while overall ambient lead concentrations decreased 95 % since 1977, and 75% since 1987.[6]

Sulfur or Oxides
Health and Welfare Effects

Sulfur, phosphorous, carbon, oxygen, hydrogen, and nitrogen are the main components of most living things, and consequently are found in fossilized forms including the fossil fuels such as coal and oil. The combustion of fossil fuels results in the oxidation of sulfur to produce sulfur dioxides, which react with moisture, oxidizing agents, and reactive hydrocarbons and other substances in the atmosphere to produce sulfates, sulfites, and sulfuric acid. The primary source of emissions are electric utilities. These compounds may produce a variety of adverse health effects. Sulfur oxides are also involved in corroding metals; disintegrating marble, limestone, and dolomite; attacking fabrics, reducing visibility; and impairing the growth of plants and forests. The most prominent health concerns associated with SO_2 include respiratory illness, effects on breathing, a reduction in lung defenses, and aggravation of existing cardiovascular disease. Persons with asthma, cardiovascular disease, and chronic lung disease are most sensitive to the effects of sulfur dioxide. Children and the elderly are also at increased health risks from the inhalation of sulfur oxides. Sulfur oxides are among the main precursors to acid deposition, with nitrogen oxides being the second greatest contribution.

Acid Deposition

Acid deposition is a term that represents a more inclusive understanding of the acidification process than the term acid rain, since acidity may be found in rain, sleet, snow, fog, clouds, and adsorbed to particles. Consequently, the term acid rain is being replaced by the term acid deposition although the two are presently used synonymously. The problems associated with acid deposition were first recognized about 30 years ago in Sweden when changes in the plant and animal life of freshwater lakes were associated with increased acidity from acidic deposition.[17] The revelation of acidic deposition was not widely accepted by the world scientific community initially, but as measurements of rainfall pH demonstrated greatly expanding areas of increasing acidity, the problem became the subject of domestic and international debates over acid rain controls. The United States passed the 1990 CAAA with major provisions for sulfur and nitrogen oxide controls, although atmospheric acidity still presents a major environmental problem.

Rain, snow, clouds, and sleet are normally somewhat acidic because of the presence of CO_2 in the atmosphere producing carbonic acid with pH levels of nearly 5.0 in the absence of pollutants. The pH scale is logarithmic and is based on the negative log of the hydrogen ion concentration. Consequently, a pH of 3.0, as has been found in precipitation in the Ohio Valley, is 100 times the acidity of rain with a pH of 5.0. About 70% of sulfur oxide emissions originate from large electric utilities burning high sulfur coal while other major contributors to NOx and SOx include industrial boilers, metal smelters, and automobiles.[17] Sulfur acids have contributed about 65% to the acid deposition problem while nitrogen oxides contributed 35%. However, as regulatory controls become more effective, the levels of sulfur oxides in the atmosphere have decreased by 39% since 1970 while ambient concentrations of NO_2 increased eight percent over the same period.[13] Consequently, the emissions of NOx from automobiles and electric power generators is taking on greater importance.

Effects of Acid Deposition on Ecology

Acid deposition is a product of fossil fuels with most SOx occurring from electric utilities burning coal or oil. Many of these plants employ the strategy of building tall stacks so the discharged gases will be carried out of the vicinity. Although locally effective, the gas streams are carried by prevailing winds great distances to be deposited far from the origin in a process known as long distance transport. Heavily industrialized areas of the Ohio River Valley, upper Midwest, and areas of Illinois, Pennsylvania, West Virginia, and Indiana are often considered to be the source of significant acid plumes which travel eastward to deposit all along the Northeast coast off the United States and parts of Canada. The heavily industrialized areas of Great Britain and Central Europe and parts of China create similar conditions for polluting distant locations. Sweden and

Norway point to central Europe as a major cause of the acidity falling on them. Although long distance transport contributes significantly to acidic deposition, local sources are also important contributors and just how much each contributes to the problem has been strongly contested by those allegedly producing the pollutants, and those who receive them.

The New England areas including upstate New York, the mountainous areas of West Virginia, Pennsylvania, Virginia, Kentucky, Wisconsin, Minnesota, parts of Canada, and western mountainous areas of North America are sensitive to acidity because these areas are deficient in magnesium and calcium carbonates. Therefore, the soils and water have low acid neutralizing capacity and are referred to as acid-sensitive ecosystem (Fig. 11-2).[15]

Aquatic Ecosystems

When lakes and streams receive acid deposition in acid-sensitive areas they are unable to neutralize or buffer the acidity. Consequently aquatic systems in the Adirondacks or New York, New England, and other sensitive areas become chronically and progressively acidified or receive sudden massive doses of acidity in spring thaws resulting in a process known as shock loading. As the acidity of a body of water increases there is a reduction in the diversity of species and a shift in species composition. The smallest of the organisms, unicellular phytoplankton, disappear followed by the benthic invertebrates. These organisms serve as important food sources and nutrient recyclers (decomposers) so that large aquatic creatures are adversely affected by their disappearance. Acidification can lead to mobilization of aluminum and possibly other toxic metals, which in combination with the increased acidity, leads to reductions in fish populations. At pH levels below 5.5, fish populations decline with the most sensitive, juvenile fish first disappearing. The eggs, larvae, and juvenile fish fail to thrive in a process called recruitment failure, often leaving acidified streams and lakes with a diminishing population of older, larger fish. Eventually, entire lakes and streams become devoid of fish, such as in the Adirondacks, mid-Atlantic mountains, New England, Ontario, Quebec, Sweden, and Norway.

Effects on Forest and Plants

The causal role of acid deposition in the decline of forested areas in North America, Germany, and Central Europe has been a matter of considerable investigation and debate. Acidic deposition at levels of pH 4.0 to 5.0, which are most common, do not appear to cause widespread adverse effects on forest ecosystems.[15] However, conifer forests such as Red Spruce on mountain tops in New Hampshire, Vermont, and in the Appalachians have been more than 80 percent decimated at the cloud line. Severe damage is also evidenced in Central Europe in such places as the Czech Republic and Poland where 60 to 70 percent of forests show evidence of damage associated with sulfur and nitrogen disposition. The mechanisms to such destruction are not immediately devious but may be attributed highly to combinations of ozone and acid clouds (as low as pH 2.2) which: (1) directly damage leaves; (2) mobilize toxic metals in soil such as aluminum, which adversely effect roots; (3) leach nutrients from soil; and (4) over-stim-

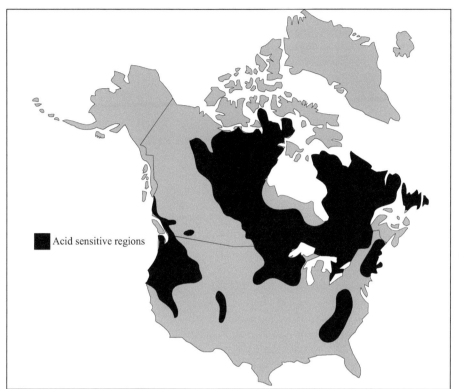

Figure 11-2. Acid sensitive regions of North America and Canada (Adapted from Godish[4]).

ulate plants from excess nitrates, which aggravates deficiencies of other nutrients. These factors may combine to increase forest susceptibility to insect and fungal pathogens.[16]

Current Directions in SOx Control

Ambient concentration of SO_2 decreased 37 percent since 1987 while emissions decreased by more than 14 percent. Reductions in SO_2 emissions are due mostly to controls implemented under the USEPA's Acid Rain Program.[13] This program features a number of options including: (1) switching to low sulfur fuel; (2) using scrubbers that remove SO_2 from the stack emissions; (3) washing coal, which removes up to 50% of the sulfur content; and (4) using advanced combustion technologies such as fluidized-bed combustion to remove sulfur and nitrogen oxides.

Nitrogen Oxides

Nitrogen and oxygen are combined in the high temperature confines of industrial boilers and automotive engines to produce nitric oxide. This is oxidized in the atmosphere to produce nitrogen dioxide, an orange-yellow to reddish-brown gas that can often be seen from a distance enveloping cities. Nitrogen oxides are an important pre-cursor for ozone and acidic deposition (nitric acid). Nitrates arising from NOx emissions fall from the sky to stimulate algae growth (i.e., red tide) causing toxic conditions for aquatic life in bays and estuaries. Nitrogenous compounds also contribute significantly to the formation of fine particulates. Nitrogen dioxide irritates lung tissues and reduces resistance to respiratory infections, while possibly increasing the incidence of acute respiratory disease in children.[13]

THE HISTORY OF AIR POLLUTION CONTROL IN THE U.S.

The control of air pollution was seen as a local control issue up until the 1950s when a federal role began to evolve. It was in 1955 that Congress authorized the Public Health Service in the Department of Health, Education, and Welfare (DHEW) to research air pollution and conduct training programs. However, air quality continued to grow worse and Congress passed the Clean Air Act of 1963, which gave federal authority to (1) develop air quality criteria for protecting the public health; (2) conduct research on sulfur dioxide pollution; (3) provide grants to establish state air pollution control agencies; and (4) provide more research, training, and technical assistance with regard to air pollution. Amendments to this Act in 1965 established the National Air Pollution Control Agency (NAPCA), and set emission standards for 1968 model light-auto motor vehicles. In spite of these regulatory initiatives, air pollution continued to worsen, and the 1963 Clear Air Act proved inadequate to deal with the problem. The response of Congress was to pass the Comprehensive Air Quality Act of 1967 which represented the first attempt to develop a regional approach for the control of air pollution through the designation of Air Quality Control Regions (AQCRs). However, states still kept the main responsibility for enforcement and were expected to develop air quality standards and plans for implementing them. Progress proved to be slow, and as the Americans celebrated "Earth Day" for the first time on April 22, 1970, public attention to environmental concerns reached a high level, stimulating Congress to strengthen air pollution control initiatives. This time Congress passed the 1970 Clean Air Act Amendments (CAAA) which: (1) dissolved NAPCA; (2) transferred air pollution control activities to the U.S. Environmental Protection Agency (USEPA) which the President created by executive order; (3) increased federal enforcement authority; (4) established uniform National Ambient Air Quality Standards (NAAQS) for 6 criteria pollutants; (5) required states to develop plans for implementing programs to achieve NAAQS in the states and submit the State Implementation Plans (SIPs) to the USEPA; and (6) gave American citizens the right to sue private and government entities to enforce air pollution requirements. Additional amendments were made in 1977 to the 1970 CAAA that postponed and extended many compliance deadlines for air-quality standards and auto emission standards.[17]

However, the 1977 CAAA did authorize the USEPA to regulate chemicals suspected of destroying stratospheric ozone, and also provided for the protection of clean air areas by preventing further air pollution through the concept of PSD or Prevention of Significant Deterioration.[17] Despite these efforts and some success under pollution control, three major threats to the nation's environments and to the health of millions of

Americans defied these regulatory initiatives. These threats included acid rain, toxic air emissions, and urban air pollution. Additionally, the threats to global environment persisted in the form of global warming by greenhouse gases, and the continued destruction of stratospheric ozone by ozone-depleting chemicals. President George Bush proposed major revisions to the Clean Air Act in June of 1989 which were passed by Congress and signed into law on November 15, 1990. These revisions are known as the Clean Air Act Amendments of 1990. Many of the provisions of this Act are still coming online as there are many progressive and creative new directions included in the Amendments.[18] There are several titles to the Act which are intended to produce a healthy, productive environment combined with a sustainable energy policy and economic growth.

Titles of the Clean Air Act
Title I: Provisions for Attainment and Maintenance of the NAAQS
Building upon the 1970 and 1977 CAAA, the 1990 CAAA attempt to strengthen the provisions protecting the public against seven of the most widespread and common pollutants designated as criteria pollutants (Table 10-1). The criteria pollutants are those for which standards have been set establishing maximum concentrations allowed in the ambient or outdoor air environment and are known as National Ambient Air Quality Standards (NAAQS). The USEPA promulgated standards for six classes of pollutants in 1971, and an air-quality standard for lead was promulgated in 1978. These standards are based upon Air Quality Criteria, which documents the relationship of these pollutants to effects on health and welfare, therefore providing a scientific basis for the standards.[17]

Over 100 million Americans live in cities that fail to meet the NAAQS for ozone, and many people continue to be exposed to unacceptable levels of carbon monoxide and particulates. The 1990 CAAA established these as non-attainment areas, which were first described in the 1970 CAAA and refers to those Air Quality Control Regions (AQCRs) as designated by the USEPA that exceed the levels of one or more of the criteria pollutants during the year. The USEPA ranked these areas according to the severity of an area's air pollution problem (USEPA). A non-attainment status carries serious implications because new industries or sources that might contribute to the pollution would be prohibited and therefore limit economic development. This was economically very confining. The USEPA, therefore, developed an offset policy which allowed an industry to build a new facility if it achieved agreements with nearby sources to reduce their emissions so no net increase in pollution levels occurred. Alternatively, new sources could be developed in a non-attainment region if the pollution emission rates could be reduced maximally by the best available control technology (BACT) as designated by the USEPA. Under the 1990 CAAA, the USEPA classified non-attainment areas for ozone, carbon monoxide, and particulates based on the extent to which the NAAQS is exceeded. The 1990 CAAA further establishes specific pollution controls and attainment dates for each control. Carbon monoxide and PM10 are classified as either moderate or serious non-attainment areas, while ozone has five non-attainment classifications ranging from marginal (Albany, NY) to extreme (Los Angeles, CA). The more severe the non-attainment problems, the more controls the area needs to apply, and the more time it is allotted to do so.

Title II: Provisions Relating to Mobile Sources

Most automobiles emit up to 80% less pollutants than in 1960, but they still account for the greatest combined amount of criteria pollutants including carbon monoxide, hydrocarbons, and nitrogen-oxides. The main reason for this is the proliferation of vehicles on the nations highways. More than one in every five households in the United States have three or more cars and the miles traveled are projected to grow by two percent each year through 2010 as Americans no longer just travel from home to work, but link their trips to shopping, after-school activities, day-care, and many other activities. Efforts by government agencies to discourage the use of single-occupant vehicles (SOVs) has been generally discouraging since Americans value their travel independence and continue to enjoy the luxury of inexpensive fuel, affordable cars, and free highways.[19] To combat this problem, the 1990 CAAA reduced the allowable emissions for carbon monoxide, hydrocarbons, and nitrogen-oxides from vehicle tailpipes through tighter inspection and maintenance (I/M) programs. The amount of evaporation of gasoline while refueling at the pumps has also been reduced through the use of specially designed recapture nozzles. Programs requiring cleaner reformulated gasoline and/or oxygenated fuels are also being phased in throughout areas of the country in non-attainment areas for ozone and carbon monoxide, respectively.

Methyl-t-butyl ether (MTBE) is the biggest volume oxygenate used in gasoline and contains at least the 2.7% O^2 by weight required to reduce CO emissions. The additives must be used during the winter months of November through December when levels of CO peak. The amount of MTBE has increased from the 1.8 billion gallons used in 1992 because the program to use reformulated gasoline to combat ozone depletion began in 1995.[20] The law also required California to phase-in the use of 150,000 clean fuel cars (i.e., propane or electric cars) by 1996 and 300,000 by the year 1999. Additionally, 26 of the country's most polluted areas will have to further regulate companies and agencies with centrally-fueled fleets of 10 or more vehicles such that vehicle emissions are further reduced. This process began in 1998.

Title III: Air Toxics

Congress recognized that very hazardous chemicals, other than the criteria pollutants were being emitted into the nations air, and directed the USEPA under the 1970 CAAA to establish a list of such chemicals along with standards and control techniques. Unfortunately, the USEPA acted very slowly over 20 years, and listed only eight hazardous air pollutants (HAPs) including arsenic, asbestos, benzine, beryllium, mercury, radionuclides, vinyl chloride, and emissions from coke ovens. Congress was aware that more than 2.7 billion pounds of toxic air pollutants were being emitted annually in the United States and that exposure to such substances may be causing up to 3000 cancer deaths annually.[18] They obtained the information that industry was required to supply under the Superfund "Right-to-Know" rule (SARA, Section 313). This rule was passed in 1986 in response to the deaths of nearly 2000 people in 1984 at Bhopal, India, because of the accidental release of toxic chemicals from a chemical manufacturing facility. This heightened concern prompted Congress to list 189 toxic air pollutants for which emissions must be reduced. The USEPA was required to publish a list of source categories that omit certain levels of these pollutants and to issue standards based on the best demonstrated control technology within the regulated industry, known as the Maximum Achievable Control Technology (MACT). The source categories are to be controlled by the year 2000, although companies voluntarily reducing emissions may obtain a six-year extension.

Title IV: Acid Deposition Control

Emissions of nitrogen and sulfur oxides are partially converted in the atmosphere to nitric and sulfuric acids, which return to the earth in rain, snow, fog, and on dry particles to cause acidic deposition. Such acidic deposition harms non-buffered lakes and streams, by killing fish and other aquatic organisms, causes forest destruction, corrodes steel, damages concrete and marble structures, reduces visibility, and even contributes to adverse health effects.[18] Despite progress in reducing emissions under the 1970 CAAA, nearly 20 million tons of SO_2 were being emitted annually in 1990, mostly from combustion of fossil fuels by electrical generating facilities. The goal of the 1990 CAAA is to reduce the total SO_2 emissions by half to 10,000 tons per year through a phased program requiring power plants to reduce emissions. The 1990 CAAA is the first air regulation to encourage the use of market-based principles, such as emission banking and trading through the use of allowances. It works as follows: coal-burning electric utilities and other industries will be allowed to emit nearly 9 million tons of SO_2, while close to 250,000 theoretical tons were set aside for auction. If a utility expected it could not reduce its SO_2 emissions to meet the new federal standard, it could purchase an "allowance" from other utilities whose emissions were lower than that required under federal regulations. An allowance is defined under 1990 CAAA as the right to emit one ton of sulfur dioxide. This is a market incentive approach where allowance, or the right to pollute, could be traded or banked on the market. This concept was originally proposed by members of the Environmental Defense Fund (**EDF**) as part of the effort to reduce SO_2 emissions 50% by the year 2010. According to EDF representatives, the program has been very successful, with emissions reductions of 3.4 million tons in one year alone at unexpectedly low program costs.[15] However, the market for the sale of allowances has significantly diminished, and auction prices for a one-ton allowance has dropped from the 1990 expected price of $1,000, to $250 in 1992, $140 in 1995, and $68 in 1996. Such inexpensive prices worry some people that utilities will find it less expensive to buy allowances and so pollute considerably more.[21]

Title V: Permits

The 1990 CAAA also introduced an operating permits program similar to that used on other federal legislation designed to protect water, such as the Federal National Pollutant Discharge Elimination System (NPDES). This updated the Clean Air Act to make it more consistent with other federal statutes. Under the permit program, air pollution sources regulated by the Act must obtain an operating permit through their state environmental program: state programs must be approved by the USEPA if they are to administer the permit system. The permit requires that all rules and regulations regarding air pollution control be listed on the one permit, making enforcement a much easier success. Additionally, the permit will list the permissible levels of pollutants and the required control measures. The source is required to file periodic reports showing its level of compliance with the permit obligations. They must also pay a fee for the permit which is intended to cover costs by the state in administering the program.

Title VI: Stratospheric Ozone and Global Climate Protection

Stratospheric ozone is most concentrated in the region of 24 to 40 km, although it is among the least concentrated of naturally occurring atmospheric gases. Ozone is formed when ultraviolet rays split oxygen molecules to release an oxygen atom that combines with an oxygen molecule to form ozone (O_3). Although created by UV light, ozone also absorbs UV energy in the range of 200 to 300 nm, protecting plant and animal life from the harmful effects of intense ultraviolet radiation. In the mid-1970s, concern about the integrity of the ozone layer grew when the scientists, Rowland and Molina from the University of California, reported that chlorofluorocarbons (**CFCs**) were attacking and destroying the ozone layer. The CFCs were widely used as refrigerants, aerosol propellants, and foam-blowing agents in styrofoam containers. The scientists began their work in the 1970s and showed that CFCs could be photolytically destroyed in the atmosphere to release chlorine atoms that could then catalytically destroy O_3 molecules in the stratosphere   (Fig. 11-3).   Mario Molina, Sherwood Rowland, and Max Planck received the Nobel Prize for chemistry in 1995 for their work in establishing that CFCs were destroying the ozone layer.[22] The work of Rowland and Molina erupted from the narrow view of a few esoteric scientists to international attention in 1985 when British scientists discovered a large hole in the ozone over the Antarctic during September to mid-October. This was subsequently confirmed by the National Aeronautics and Space Administration (NASA).[14] The Antarctic hole has continued to grow since then, while also lasting longer. In the early 1990s, NASA scientists reported that very low concentrations of stratospheric ozone were appearing over the middle latitudes of 20 to 60 degrees. While some ozone loss was partially due to volcanic dust and gases hurled into the atmosphere by the eruption of Mount Pinatubo in the Philippines, losses in ozone were being observed over the United States, Japan, Europe, Russia, and China before and since then. The World Meteorological Organization reports the ozone hole over Antarctica as being approximately as large as the combined area of Canada and the U.S., peaking at 7.7 million square miles and lasting for 50 days. Scientists predict the ozone will likely grow larger over the next 15 to 20 years. Despite global efforts to halt the production of ozone-depleting CFCs, there remain substantial amounts of CFCs to be released, and CFCs may persist in the stratosphere for many years.[23]

The efforts to control and halt ozone-depleting chemicals has been controversial. However, representatives from 29 nations met in Montreal, Canada, and reached international agreement (Montreal Protocol) to freeze CFC production at 1988 levels by

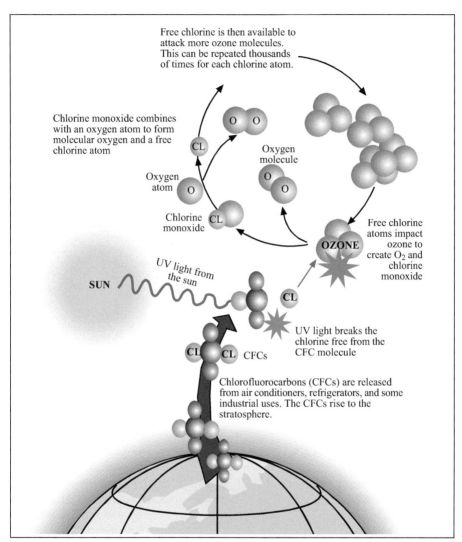

Figure 11-3. The catalytic destruction of ozone in the atmosphere by CFCs according to the Rowland-Molina hypothesis.

1990, and then to cut production in half by the year 2000. Mounting evidence that the ozone layer was being depleted resulted in a subsequent conference in London in 1990 and the Montreal Protocol was amended to hasten the phase-out of CFC production by the year 2000. The amendments also called for a phase-out of ozone-depleting halons (used in fire extinguishers) and CCL4 by the year 2000. Methyl chloroform is to be phased out completely by the year 2005. More than 80 countries signed the 1990 agreement in London. The 1990 CAAA required that Class I ozone-depleting chemicals be phased out on a schedule specified in the 1987 Montreal Protocol. Class II chemicals

(HCFCs) were to be phased out by 2030. However, as stratospheric ozone depletion became progressively worse, and became apparent over North America and northern Europe, representatives of more than 100 nations met in Denmark and agreed to speed up the phase-out of CFC production by January 1, 1996, and Halon by 1994. [17]

The efforts to find substitute chemicals has usually resulted in inferior performance and the requirement to modify existing equipment to make use of the alternatives such as Hydrofluorocarbons (HCFCs) or chlorine-free HFC-134a (CF3CFH2). These substitutes are not expected to damage the ozone layer, but are still infrared absorbers and contribute to the greenhouse effect. Therefore, the amendments to the Montreal Protocol target many CFC substitutes for phase-out by 2040 through voluntary efforts.[17] However, there was concern that many LDCs would be unable to afford the expense of modifying equipment to accept the substitute refrigerants. The Montreal Protocol allowed for continued production of CFCs in developed nations for use in developing countries in order to prevent them from building their own CFC-producing factories. American factories may produce 53.500 tons each year for export until 2005.

Unfortunately, there has been evidence that CFCs have been returning to this country illegally where a 330-pound cylinder that costs $70 in Europe is sold for $242 in the U.S.. This is still less costly than the CFCs recycled in the U.S. Officials believe as much as 10,000 tons entered the U.S. in 1995 from Russia and other countries in the former Soviet Union.[24,25] Despite these troublesome and controversial events, overall production of CFCs is down markedly and the National Oceanic and Atmospheric Administration (NOPA) has reported a reduction in the rate at which CFCs are accumulating in the atmosphere.[26] The minimum value of 88 Dobson units was recorded in 1994.[27] Although there appears to be a slight increase in stratospheric ozone compared to 1994 the recovery of stratospheric ozone is expected to take decades or up to a 100 years.[26]

Other Titles to the 1990 CAAA

The Clean Air Act Amendments of 1990 contains many new provisions with regard to tougher enforcement, new research, and development in air monitoring. There are also unemployment benefits through the Job Training Partnership Act for those workers who may have lost their jobs because industries had to comply with the new provisions of the 1990 CAAA. Efforts under the Act also attempt to reduce the problem of pollution-created haze areas at National Parks and other parts of the country.

Revised Ozone and Particulate Standards

On July 17, 1997, the USEPA announced new NAAQS for ground-level ozone and particulates. Ground-level ozone is a major component of smog that is photochemically produced as a secondary pollutant of the stratosphere from the interaction of sunlight, nitrogen oxides, and hydrocarbons. The previous ozone standard was last revised in 1979 and set at 0.12 ppm for one hour. However, based on more than 3000 new studies on the ecological and health effects of ozone published since 1980, its become clear that the standard did not adequately protect the public health, and the USEPA replaced the previous standard with an 8-hour standard set at 0.08 ppm.[27]

Similarly, the USEPA announced a new standard for particulate matter (PM) under the national ambient air quality standards (NAAQS) based on the review of hundreds of scientific studies. USEPA has added a new annual PM 2.5 standard set at 15 micrograms per cubic meter ($\mu g/m^3$) and a new 24-hour PM 2.5 standard set at 65 $\mu g/m^3$. The USEPA is retaining the current annual PM10 standard set at 50 $\mu g/m$.[3, 9] These changes to the standard were met with significant opposition, from businesses such as the auto industry, oil corporations, and from some state governors. Areas of the country producing more smog and soot than considered healthy under the proposed standards would face economically-strangling restrictions including controls on truck operations, mandates to reduce pollution-causing auto emissions to zero, and requirements that vehicles be run on fuels other than gasoline, strict controls on power plants and factories, and increased emission checks to autos. The new standards are considerably tougher than the ones developed under the 1970 CAAA, and 1990 CAAA which are still unmet by many states across the country.[28]

THE ISSUE OF GLOBAL WARMING
The Hot Air Treaty, Kyoto, Japan

The global warming treaty completed in December 1997 (Kyoto, Japan) asked Western nations to reduce greenhouse gases to pre-1990 levels by 2010. The U.S., if it were to comply, would need to reduce greenhouse emissions to seven percent below the level of the year 1990 or by about 33 percent, one third of its otherwise projected use in the year 2010.[29] Western European nations and Japan are assigned similar goals. According to the agreement, reductions would not start taking effect until 2008. Even if the nations complied, total world production of CO_2 would be reduced by only 1.1 billion tons by the year 2010. None of the developing nations have agreed to the Kyoto Treaty, even though they are soon expected to produce more greenhouse gases than the West. The LDCs have the urgent problem of providing for exponentially increasing populations, and further reducing use of fossil fuels is unlikely. Inexpensive labor and unregulated industry could attract business from the West thereby shifting jobs, machinery, and CO_2 production to the LDCs with no net overall decrease in greenhouse gases, while greatly disturbing Western economies. The U.S. Congress has already voted down any treaty that does not include the LDCs, and if the accord is not ratified, global emissions of greenhouse gases could become double their pre-industrial level by the year 2050.[29] As an example, China, with its heavy reliance on coal, will be the largest single producer of climate warming carbon dioxide by 2010 to 2030. Despite rising health problems among its citizens because of increased air pollution, there appears to be no slowing in China's coal burning practices as it races to feed its billions of people and enter the world's industrial market place.[30]

REFERENCES

1. **Chamber. L.,** *Classification and Extent of Air Pollution*, Stern, A.C., ed., 3rd ed., Academic Press, New York, 1976., Chapt. 1.

2. **Morgan, M.T., Gordon, L., Walker, B. et al.,** *Environmental Health*, 2nd ed., Morton Publishing Co., Englewood, CO, 1997.

3. **Whelan, E.M.,** *Toxic Terror,* Prometheus Books, Buffalo, New York, 1993, Chapt. 1.

4. **Godish, T.,** *Air Quality*, 3rd ed., CRC/Lewis Publishers, Boca raton, FL, 1997, Chapt. 5.

5. **Liu, B. and Xu, E.,** *Physical and Economic Damge Functions for Air Pollutants by Receptors*, USEPA/600/6-76-011, 1976.

6. **USEPA,** *National Air Quality and Emissions Trends Report,* 1996, USEPA Office of Air Quality Planning & Standards, USEPA document 454/R-97-013 1/13/98. <wysiwyg://http:www.usepa.gov/oar/aqtmd96/trendsfs.html>

7. **Allen, S.,** *More dying of dirty air than in cars, study finds,* Boston Globe, May 9, 14, 1996.

8. **Washington (CNN),** *America's dirty little secret: smog still a health problem*, CNN Interactive, June 21, 1996.

9. **USEPA,** *USEPA's Revised Particulate Standard, FACT sheet.* Office of Air Quality Planning & Standards, July 17, 1997. <http://ttnwww.rtpnc.usepa.gov>

10. **Godish, T.,** *Air Quality*, 3rd ed., CRC/Lewis Publishers, Boca Raton, FL, 1997, Chapt. 2.

11. **USEPA,** *USEPA's Revised Particulate Standard, FACT sheet.* USEPA Office of Air & Radiation, Office of Air Quality Planning & Standards, July 17, 1997. <http://ttnwww.rtpnc.usepa.gov>

12. **Pennsylvania Department of Environmental Protection,** Appeals Court puts the brakes on EPA's 8-hour ozone standard. *DEP Update Article*, 5, 21, 1999, http://www.dep.state.pa.us/dep/deputate/airwaste/aq/standards/ozone/update_

052199.htm

13. **USEPA**, *National Air Quality and Emission Trends Report, FACT sheet*, USEPA, Document Number 454/e-97-013, USEPA Office of Air Quality Planning and Standards, 1, 23, 1998. <wysiwyg://11http:www.usepa.gov/our/aqtrnd96/trendsfs.html>

14. **Anonymous**, Carbon monoxide-heart failure link., *Env. Health Perspectives*, Vol. 104, 21, 138, 1996.

15. **Godish, T.**, *Air Quality*, 3rd ed., CRC/Lewis Publishers, Boca Raton, FL, 1997, Chapt. 4.

16. **Flynn, J.,** The following forest, *Amicus Journal*, 15, 4, 1, 1994.

17. **Godish, T.**, *Air Quality*, 3rd ed., CRC/Lewis Publishers, Boca Raton, FL, 1997, Chapt. 8.

18. **USEPA**, *The Clean Air Act Ammendments of 1990: Summary Materials*, USEPA Office of Air and Radiation, reproduced by Library of Congress, Congressionsal Research Service, November 15, 1990.

19. **Kessler, J. and Schroder, W.,** *Meeting Mobility and Air Quality Goals, Strategies That Work*, USEPA, Office of policy Analysis, October 13, 1993.

20. **Anderson, E.V.**, Health studies indicate that MTBE is safe gasoline additive, *C&EN*, September 20, 1993, 9.

21. **Bradley, J.**, Buying high, selling low, *E/The Environmental Magazine,* July/August, 1996. <http://www.emagazine.com/2curr3>

22. **Chandler, D.L.**, MIT scientist shares Nobel for identifying ozone damage, *The Boston Globe,* October 12, 1995, 101V.

23. **Associated Press (AP)**, Ozone hole grows to record size, *CNN Interactive,* November 12, 1996.

24. **Wald, M.L.,** Smuggling of polluting chemicals is resolved, *NY Times National*, 17, 1, 1995.

25. **Begley, S.,** Holes in the ozone treaty, *Newsweek*, September 25, 1995, 70.

26. **Stevens, W.K.,** Scientists report an easing in ozone-killing chemicals, *New*

York Times, August 26, 1993.

27. **USEPA,** The size and depth of the ozone hole, written by USEPA's Stratospheric Protection Division, June 10, 1998. <http://www.usepa.gov/ozone/science/hole/size.html>

28. **Gerstenzank, J.**, Clean-air plans fuels backstage U.S. fight, *Los Angeles Times*: April 26, 1997. <http://www.latimes.com/home/news/science/environ/x000035587.html>

29. **Easterbrook, G.**, Hot air treaty, *U.S. News & World Report,* December 22, 1997, 46.

30. **Tyler, P.,** China's inevitable dilemma: coal equals growth, *New York Times,* November 29, 1995, A1.

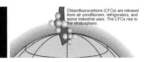

WEB LINKS FOR THIS CHAPTER

Air Pollution Emissions Overview
http://www.epa.gov/oar/oaqps/emissns.html

Air Pollution Operating Permits
http://www.epa.gov/oar/oaqps/permitupdate

Air Toxics
http://www.epa.gov/oar/airtoxic.html

EPA Acid Rain Program Home Page
http://www.epa.gov/acidrain/ardhome.html

EPA Office of Air Quality Planning and Standards
http://www.epa.gov/oar/oaqps

EPA Office of Mobile Sources
http://www.epa.gov/OMSWWW

Full Text of the CAA
http://www.epa.gov/oar/caa/contents.html

Guide to the CAA
http://www.epa.gov/oar/oaqps/peg_caa/pegcaain.html

Health Effects of Ozone, Regional Ozone Maps, Office of Air
and Radiation
http://www.epa.gov/oar/oaqps/airnow

National Emission Standards for Hazardous Air Pollutants
http://www.epa.gov/docs/epacfr40/chapt-I.info/subch-C/40P0061

Summary of the Clean Air Act
http://www.epa.gov/region5/defs/html/caa.htm

What You Should Know About Ozone
http://www.magnet.state.ma.us/dep/bwp/daqc/files/ozonefaq.htm

WATER QUALITY REGULATIONS *12*

Kate Brossard

Water is an essential component of everyday life. It is used not only for drinking, but also for irrigation, recreation, food preparation, and a habitat for plants and wildlife. Industry and human consumption have increasingly threatened the quality of waterways. For these reasons, it is important that water be protected from pollution both on the surface and through the soil. Policy has evolved from the late19th century and has since become an ever-expanding topic of regulation. Following is a summary of the most important regulations that have come to pass, beginning with a brief history and continuing through to present-day policy.

Major Sources of Pollution and Methods of Abatement

Most pollution can be pinpointed to industrial processes, municipal waste-treatment facilities, and agricultural processes. All of these are responsible for different types and amounts of pollution. The main categories of pollutants are:

Organic Wastes – measured by way of "biochemical oxygen demand" (BOD). BOD is the amount of oxygen required in the decomposition of organic material at certain temperatures and over a certain amount of time. An increased demand in oxygen for the decomposition of waste material results in a decrease in available oxygen for the existing plants and fish. Over time, a waterway can adjust to these fluctuations in small amounts, but for any prolonged period of time, lasting damage may occur. Organic wastes in the environment result largely from municipal waste-treatment plants;

Other Nutrients – result in accelerated rates of eutrophication. Eutrophication is the natural degradation of lakes and basins occurring from heightened levels of nutrients and sediments. Over time, lakes become shallower and more productive, eventually drying up. However, this can take thousands of years. By accelerating this process with the addition of detergents, fertilizers, human and animal waste, and soil runoff (usually a result of agricultural lands), lakes are disappearing at a much faster rate;

Toxic Chemicals and Other Hazardous Substances – all impair aquatic life and result in unsafe drinking water for humans. This category includes heavy metals and is often a result of industrial processes;

Heated Water – known commonly as "thermal pollution," adding hot water to waterways can speed up biological and chemical processes, resulting in less retention of dissolved oxygen. In turn, fish and plant-life have more difficulty surviving.[1]

Pollutants Can Also be Categorized More Generally

For the purposes of technology-based standards, a few terms must be introduced here. "Best Practicable Technology" (**BPT**) refers to technology that is readily available and at a relatively low cost, while "Best Available Technology" (**BAT**) is much more stringent because it includes all technologies, even if they are more costly. "Best Conventional Technology" (**BCT**) allows more lenience than BAT in that it accounts somewhat for cost (40 CFR Section 301(b);[2] full text available at <*http://www.epa.gov/epacfr40/cfr40toc.htm*>. However, these terms are fairly loose in that they do not specify specific levels of technology that can be applied to all industries. It was determined that EPA would be responsible for enforcing these standards. Firms that meet certain criteria may be allowed different levels of variation. The following categories are used for setting general technology-based standards:

Toxic Pollutants – listed by the EPA. These were required to achieve BAT by 1984 as set out by the Clean Water Act of 1977. In 1987, the deadline was extended to 1989;

Conventional Pollutants – also listed by the EPA. These include BOD, fecal matter, suspended solids, and pH. These were required to achieve BCT by 1984 as of 1977 and the deadline was also extended to 1989;

Nonconventional Pollutants – includes all other pollutants that pose a threat to human health or waterways. Likewise, firms discharging these pollutants were required to achieve BAT by 1984. This regulation was also extended to 1989 or 3 years after limits were established, whichever came later.

All of these pollutants can be slowed down or prevented altogether with the appropriate abatement plans (see 40 CFR Section 130 at <*http://www.epa.gov/epacfr40/cfr40toc.htm*>). Treatment and production processes can be altered to be more efficient. Waste can be treated in municipal waste-treatment systems in order to remove BOD and nutrients that may be harmful in large quantities. In municipal waste-treatment facilities, treatment is divided into three categories:[2]

Primary Treatment – separates solids in preparation for further treatment or disposal. Generally, 25 to 30% of BODs are removed in this process;

Secondary Treatment – a biological process that speeds up the naturally-occurring degradation process and removes up to 90 percent of BODs;

Advanced Treatment – a variety of treatments, each specific to the needs of different types of waste. Up to 99% of BODs can be removed in this process.

The problem still remaining with all of these processes is that large quantities of phosphorus and nitrogen usually remain (anywhere from 70 to 80 percent). Also, street runoff (which is difficult to control) may contain organic and toxic wastes that will inevitably enter the water system. Therefore, it is necessary to reduce pollution at the source, an issue that has been addressed over the past century by way of the following Acts.

BRIEF HISTORY

The first water quality efforts were made in 1899 when the United States government realized that degrading water quality was posing a problem for both the natural environment and the general navigational use of waterways. It was in that year that the Refuse Act was established, giving the U.S. Army Corps of Engineers the power to distribute permits for refuse disposal in waterways. It was not until 1948 that the next legislation was passed, this time focused much more on the importance of environmental protection. The Water Pollution Control Act of 1948 was implemented so that federal authorities could assist states in water quality control by investigating, researching, and surveying areas for water quality. Once this assistance was given, the states would be on their own in terms of implementation and reinforcement. Likewise, no federal authority was established to set standards or limit pollution levels. For these reasons, the Act was amended in 1956, giving states the authority to establish a system of regulations that would produce acceptable levels of water quality. It also allowed the federal government to make grants for municipal waste treatment facilities, covering up to 55% of start-up costs. This aspect of the amendments remained an important part of the water quality movement. Ambient quality standards (standards for the surrounding environment) were established under the Water Quality Act of 1965.[2]

In 1972, the most stringent regulations up to that point were established. The Federal Water Pollution Control Act (FWPCA) Amendments provided technology-based effluent standards, established and enforced by the federal government. The Act called for a goal of zero pollution discharge by 1985, in the hopes of achieving "fishable and swimmable" waters nationwide.[1] It set out to do this by establishing limits for point sources, that is, waste from industrial and wastewater treatment plants or other such identifiable sources. The Act required that such sources achieve BPT by 1977 and BAT by 1983. Since these technology-based standards were not well defined, the effects of the 1972 Amendments were slow and the EPA was forced to reevaluate the regulations once again.

Clean Water Act Amendments of 1977

Further Amendments to the FWPCA in 1977 formed the base of most regulations still in place today. Extensions were given on the technology-based effluent standards, including BPT, by 1979 (instead of 1977) if proof could be given that genuine attempts had been made to achieve the standards. If failure to meet BPT standards was due to lack of federal funding, an extension was given until 1983. Also included in the Amendments were documented standards, permits, and enforcement programs. Publicly owned treatment works (POTWs) were required to achieve secondary treatment by 1977 and BPT by 1983. New sources in specified groups were held to stricter standards and expected to achieve BAT by 1983.

Also under these Amendments, the EPA was required to create a list of toxic substances and provide limitations based on protection of public health and water quality, instead of technical feasibility. Under Section 307 of the Act, limits must be set in order to provide an "ample margin of safety".[1]

Section 402 of the Act established the National Pollutant Discharge Elimination System (NPDES), under which the EPA or states with EPA-approved programs can grant permits to point sources for the discharge of any set amount of pollution. The program requires that a permit accompany the discharge of any amount of pollutant. Also included in the program are enforceable compliance schedules needed in order to achieve the 1977 and 1983 compliance deadlines.[1] To ensure compliance, monthly discharge reports are required for submission to the EPA.[3]

NPDES permits allow for discharge of a pollutant from a point source, normally for five years at set level, under Section 402 of the Clean Water Act. This includes industrial wastewater discharge, collected stormwater runoff, surface waters, roof runoff, tidewater, subsurface drainage, cooling water, ground water, gasoline, solvents, and concentrated discharges. The permits are issued by the EPA or by a state EPA-authorized program. As of 1999, 43 states and territories had permit capability. In order to apply for a permit, the firm must provide detailed information regarding its facilities and processes. Separate applications are necessary for different groups of pollutants. Upon evaluation, the permit is granted with various conditions of agreement. These conditions may vary from monitoring requirements and discharge limits to specific technological and quality requirements for the specific pollutant. Requirements for individual pollutants can be found in the Environmental Law Handbook [4] or under 40 CFR Section 122.

Finally, the Amendments of 1977 allowed for the implementation of stricter standards by the EPA (as little as zero discharge) on particularly high discharge point sources. This was allowed in order to protect public health or to achieve ambient water quality standards by the set deadlines. Provisions were also included to monitor industries by way of inspection, entry and monitoring, EPA enforcement suits, and citizen suits. Likewise, citizen suits could be made against the EPA for failure of performing its regulatory duties. These portions of the Amendments are still in effect today.[1]

The CWA Amendments of 1977 arose partly due to the ambiguous definitions of toxic substances set out by the previous Act. The EPA needed to issue limits and new source standards for 21 industries requiring BAT. A list of toxic substances can be found in Section 307(a)(1) of the current statute <*http://www.epa.gov/epacfr40/ cfr40t0c.htm*>. Each substance is subject to limits, as stated in Section 307(a)(2). Also, pretreatment standards are set for waste going into municipal waste-treatment systems.[1]

Clean Water Act Amendments of 1987

Further Amendments were made in 1987 in order to address certain issues that needed more attention at that time. Largely, these Amendments tightened the regulations on non-point sources such as runoff from agricultural and urban areas (Section 319). States are required to identify bodies of water that cannot achieve water quality standards by regulating point sources alone. Also, they are required to establish management programs for each source. Under these Amendments, the EPA is required to apply stricter standards to toxic "hotspots" where BAT, new source, and pretreatment standards will not achieve the established water quality standards. As of 1993, the EPA had the power to administer these stricter standards in 15 states. The other 35 had to be forced by the EPA to set the standards on their own.[3] Under the requirements of the Toxics Control System, states must identify the point sources responsible for the pollution, as well as the levels of discharge. Upon determining these factors, they must devise control strategies for each source that must be completed in three years.[1] Also under these guidelines, "backsliding" was outlawed. That is to say, new permits must be at least as strict as the old ones. The latest amendments to the Clean Water Act can be found at <*http://www.epa.gov/fedrgstr/*>. Following is a program specifically intended for the regulation of non-point sources.

Coastal Zone Management Act and the Outer Continental Shelf

Sections 1451–1464 of 16 USCA (U.S. Coastal Area) outline the Coastal Zone Management Act (**CZMA**), intended to regulate non-point sources and encourage states to develop land-use plans for coastal areas. Beyond three miles from shore, the federal government is responsible for the quality of water. However, the Outer Continental Shelf Lands Act (**OSCLA**) (43 USCA 1331) was amended in 1978 to provide a leasing program for private firms through bidding. As a result, firms have use of the waters for exploration, development, production, and sale of recovered minerals. This may conflict, at times, with the interests of the CZMA. For this reason, Section 307(c) was established to maintain "federal consistency," so that all state regulations are complied with. Safeguards have also been established in order to protect natural resources in case of oil spills or other damages to the waterway (43 USCA 1334(a)(1)). See <*http://www.epa.gov/owow/nps/czmact.html*> for further information.

Industrial Wastewater and Stormwater

Industrial wastewater and stormwater are both covered under the NPDES and provide good examples of how the permit system works. Often there are additional state and local regulations that must be conformed to in addition to federal standards. Stormwater is defined as the excess water that runs into a drainage area during and after periods of rain. This may include contaminants that are found on the ground's surface, and therefore require the attention of the Clean Water Act.[4] Industrial wastewater is defined as all wastewater that is not associated with households in the way of toilet and lavatory fixtures when used for the purpose of sanitary hygiene (known as Sanitary Wastewater). Discharge limitations include but are not limited to the following:

1. Temperature greater than 104 degrees Fahrenheit (40 degrees Celsius);
2. Toxic or non-toxic solids, liquids, or gases that, in great quantity, have the potential to injure or interfere with any waste treatment process, put humans or animals at risk, cause nuisance, or exceed the set limits;
3. Any water or waste that may, on its own or in reaction to something else, emit chemical contaminants in the atmosphere;
4. Any liquid, solid, or gas that may cause fire or explosion either on its own or in reaction to something else;
5. A pH lower than 5.0 or higher than 10.0 or any other corrosive properties;
6. More than 100 ppm of petroleum oil or any other non-biodegradable oil;
7. Solid or viscous materials that might cause obstruction in sewers if present in high quantities.

The EPA issues two general permits for stormwater. A construction site permit limits stormwater runoff from construction sites while an industrial site permit covers runoff from all other industrial sites. The Amendments of the Clean Water Act in 1987 added NPDES requirements for stormwater (<*http://www.epa.gov/epacfr40/cfr40toc.htm*>). These include provisions for source pollution reduction (spill prevention, periodic inspections and testing, maintenance of pollution control equipment, keeping floors clean, and routine cleaning of equipment). If stormwater is discharged into combined sewers, the NPDES program does not apply, but pretreatment plans become a requirement. The Stormwater Pollution Prevention Plan ensures compliance with NPDES permits by identifying pollution sources and implementing control measures.

Oil Pollution Liability

The Offshore Oil Spill Pollution Fund (OISPF) was also established under the 1978 Amendments of OSCLA, establishing liability of both the vessel operator and owner. The fines are subject to caps, though these caps are not applicable if the spill occurs as a result of negligence or intentional misconduct. The Fund is supported by a three cent per barrel tax on all transported oil. Other Acts apply to local areas in addition to the OISPF.[1]

Later, in 1990, the Oil Pollution Act was established to replace Section 311 of the CWA. It requires that liability be held by all responsible parties for clean-up costs, damages to natural resources, damages to property, and lost profits due to the disruption. Responsible parties may include the owner(s) and operator(s) of the vessel, on-shore facilities, and/or pipelines. There is a Federal Oil Spill Response Fund controlled by the President, who is also responsible for establishing damage-measurement regulations.[1] Full text of the Oil Pollution Act of 1990 is available at <*http://www4.law.cornell.edu/uscode/33/ch40.html*>.

Dredge and Fill Permits

Section 404 requires permits for the discharge of dredge or fill material (i.e. for filling in swamps). Many wetlands are subject to this regulation since they are important habitats for a wide variety of species. Under this section, it is also possible to have illegally filled areas returned to their natural state in an effort to rehabilitate the environment. This is not only a recovery measure, but also an attempt to deter filling in the future.[1] Dredging and filling is permitted in certain circumstances if carried out by the U.S. Army Corps of Engineers under Section 301 of the CWA. The regulations for such permits can be found in 33 CFR 325.1 [4] at *<http://www.epa.gov/epacfr40/cfr40toc.htm>*..

Safe Drinking Water Act

Originally enacted in 1974, the Safe Drinking Water Act (SDWA) was amended in 1986 and most regulations outlined under that version are the same today. The Act requires the EPA to set maximum levels of toxic contamination in water used by the public (listed in 42 USCA 300f–300j-26). Underground sources such as pesticide leaching, waste leaks from landfills, and underground injection wells are considered problematic in at least 34 states. Regulated levels are set based on health risks, though system operators must consider economic and technological feasibility in their approaches. Operators are required to achieve BAT standards. States may assume the responsibility of regulating, monitoring, and enforcing these levels if their standards are at least as strict as federal standards.[1]

The Regulation is divided into two main sections: National Primary Drinking Water Regulations (NPDWRs) and National Secondary Drinking Water Regulations (NSDWRs). NPDWRs identify water pollutants that have proven health risks associated with them, and therefore have maximum contaminant levels (MCLs). If levels cannot be measured, then specific treatment plans apply. NSDWRs identify aesthetic problems of drinking water, such as color and odor, that are not pleasing to the consumer, but do not pose any known health effects. Federal law does not require that these be regulated, although many states have adopted their own regulation in order to abate these problems. The 1986 and 1996 Amendments to this Act have required that the EPA establish MCLs for 83 major contaminants under NPDWR, as well as treatment techniques for contaminants that cannot be measured. As a result, each contaminant under the Primary Regulation must have the following:

1. MCLG (maximum contaminant level goal);
2. MCL (maximum contaminant level) or specified treatment method;
3. Monitoring requirements (location, frequency, and method of sampling);
4. Methods for evaluation of samples including MDL (maximum detection limit) and PQL (practical quantification level);
5. BAT as a recommendation only;
6. Time-limits on reporting and record retainment;
7. Public notice of any violation of NPDWR.

Since pesticides are a particularly important issue, the SDWA and Federal Insecticide, Fungicide, and Rodenticide Act (FIFRA) both have provisions for the protection of drinking water. Pesticides are defined as any substances intended to prevent, kill, repel, or mitigate any pest or plant. All pesticides that are to be sold or used commercially must be registered through FIFRA and approved by the EPA. The Act also has provisions for labeling and unreasonable side effects.

The Act also includes provisions for the accurate labeling of bottled water, as well as regulations pertaining to mineral and contaminant levels. These levels are at least as strict as those set out by the NPDWRs. Similar regulations apply to water coolers, in addition to maximum lead levels.[4] The most current information about this Act can be found at <*http://www.epa.gov/safewater/sdwa25/sdwa.html*>.

Underground Injection Control Programs

Underground injection of industrial waste is slowly being phased out by the Resource Conservation and Recovery Act. However, it remains an important issue as part of the Safe Drinking Water Act. The Underground Injection Control Program is a permit system that allows industries to dispose of their waste, while at the same time ensuring that all local water wells, septic systems, and the water table are not at risk. States are required to implement these programs, as regulated under Section 1423 of the Safe Drinking Water Act,[4] at
<*http://www4.law.cornell.edu/uscode/42/300f.html*>.

Federal Contingency Plans for Toxic Spills (Notification and Response)

Any non-transportation-related facility that transports or stores oil or is located within a hazardous distance of navigable waters or shorelines must be certified under the Oil Pollution Act (**OPA**) of 1990. Also, Facility Response Plans must be present if:

1. A facility transfers oil over water to or from a facility that can hold at least 42,000 gallons of oil; or
2. A facility with total oil storage of at least 1,000,000 gallons either does not have a secondary containment for its largest tank, is located within a hazardous distance of fish or wildlife, is located within a hazardous distance of a public drinking water source, or has had a reportable spill of at least 10,000 gallons in the past five years.

Other facilities may also be eligible for the facility response plan requirement depending on transportation methods, proximity to wildlife and drinking water sources, storage capacity, spill history, and other such factors.

Immediate spill notification is obligatory if any waterways are contacted or if the spill is of large quantity. Reports should be made to the National Response Center. A quantity of oil that is considered hazardous is any amount that causes sheen on the surface of water or sludge beneath the surface. More than one reportable release within 24 hours requires a call to the National Response Center.

All owners and operators of the facility causing a spill are responsible. They may

be held accountable in one or more of the following ways: penalties, clean-up costs, and/or compensation of damages to natural resources. Liability is often so high that it is more economical to institute prevention plans and required equipment than to pay fines, get sued, or receive bad publicity due to noncompliance. Charges and penalties are outlined in Section 311 of the CWA at
<http://www.epa.gov/epacfr40/cfr40toc.htm>.

Enforcement

All laws and regulations are enforceable by the EPA, and many are enforceable by state and local programs. More often than not, especially in the case of large catastrophes, the federal agency takes responsibility for action. If a state does take its own action, penalties are not required to be as high as federal standards. Since most pollution and technology levels are reported by the actual facilities, regulation is fairly simple in comparison to other federal laws.[4] However, it is necessary to monitor all industries to ensure that they are being honest. Also, there is much litigation over special exceptions to the laws as well as lawsuits filed by citizens and the EPA itself. Therefore, the task of achieving clean waters is an ongoing progression. For more information on why clean water is so important for the environment and the economy, visit the EPA website at *<http://www.epa.gov/OWOW/NPS/index.html >*.

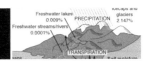

REFERENCES

1. **Findley, R.W. and Farber, D.A.**, Environmental Law in a Nutshell, *4th ed.*, West Publishing Co.,St. Paul, MN, 1996.

2. **Field, B.C.,** Environmental Economics: An Introduction., New York: The McGraw-Hill Companies, New York, 1997

3. **Freedman, M. and Jaggi, B.,** Air and Water Pollution Regulation., Westport, CT: Quorum Books, 1993.

4. **Sullivan, T. F. P.**, ed., Environmental Law Handbook, 15th ed. Rockville, MD, Government Institutes, 1999.

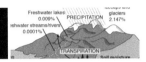

WEB LINKS FOR THIS CHAPTER

Drinking Water and Ground Water Protection Programs
http://www.epa.gov/OGWDW

EPA Ground Water Rule
http://www.epa.gov/OGWDW/gwr.html

Full Text of the CWA
http://www4.law.cornell.edu/uscode/33/ch26.html

Full Text of the SDWA
http://www4.law.cornell.edu/uscode/42/300f.html

Map of States with Approved Comprehensive State Ground
Water Protection Programs
http://www.epa.gov/OGWDW/csgwppnp.html

Massachusetts Clean Water Act
http://www.state.ma.us/legis/laws/mgl/21-26.htm

National Stormwater Center
http://www.gen.com/storm-water

Nonpoint Source Pollution Control
http://www.epa.gov/OWOW/NPS/index.html

Summary of The Clean Water Act (CWA)
http://www.epa.gov/region5/defs/html/cwa.htm

Summary of the Safe Drinking Water Act (SDWA)
http://www.epa.gov/region5/defs/html/sdwa.htm

Wastewater Management
http://www.epa.gov/OWM

ENVIRONMENTAL JUSTICE

13

David Gillum

THE BEGINNING: Executive Order 12898

In February 1994, President William Jefferson Clinton issued Executive Order 12898, entitled *Federal Actions to Address Environmental Justice in Minority Populations and Low-Income Populations.* President Clinton stated that "Each Federal agency shall make achieving environmental justice part of its mission by identifying and addressing, as appropriate, disproportionately high and adverse human health or environmental effects of its programs, policies, and activities on minority populations and low-income populations."[1] (To see the entire Executive Summary, visit *<http://www.nap.edu>.*) The signing of the Order coincided with the *Symposium on Health Research and Needs to Ensure Environmental Justice*, sponsored by the National Institute of Environmental Health Sciences in conjunction with the Agency for Toxic Substances and Disease Registry. Ultimately, the proceedings of that symposium led to a request that the Institute of Medicine produce a report with recommendations on the research, clinical, and educational needs required to achieve environmental justice.[2]

The *Symposium on Health Research and Needs to Ensure Environmental Justice* recommended that the federal government and other environmental justice stakeholders empower communities, increasing their self-determination and enhancing their interaction, collaboration, and communication with a more responsive, reinvented government. The Symposium asked for specific requests of the government to:[3]

- conduct meaningful health research;
- promote disease prevention and pollution strategies;
- set up new ways for interagency coordination;
- provide outreach, education, and communication; and
- design legislative and legal remedies to environmental justice concerns.[3]

The United States Environmental Protection Agency (U.S. EPA) defines *environmental justice* as "the fair treatment and meaningful involvement of all people regardless of race, ethnicity, income, national origin, or educational level with respect to the development, implementation, and enforcement of environmental laws, regulations, and policies." [1] <*http://www.nap.edu*> Environmental justice was mainly brought into public forum by environmentalists and Civil Rights activists, working to ensure the rights of low-income and minority communities to *clean and healthy environments.*

"Environmental Justice" as a political movement had its beginnings in a small, predominantly African American community in the southern United States. Although there had been awareness of the disproportionate burden placed on minorities and low-income communities, events did not give rise to this "movement" until 1982 in Warren County, North Carolina.[4] (Visit the web site of the Thurgood Marshall School of Law, Environmental Justice Clinic:<*http://www.tsulaw.edu/environ/envhist.htm*>).

North Carolina, 1982

As land is developed, less space is available for waste disposal, and designating new areas for landfills and hazardous waste sites has become a very difficult task. In the early 1980s, the state of North Carolina chose to build a toxic waste landfill in an overwhelmingly low-income and minority community in Warren County. This landfill was created for the disposal of PCB (polychlorinated biphenyl) contaminated soil, removed from 14 counties throughout the state.[4]

See <*http://www.tsulaw.edu/environ/envhist.htm*> for more information. The selection of this landfill sparked widespread protests and marches, which led to the arrest of more than 500 people. Among those arrested were Dr. Benjamin F. Chavis, Jr. (Executive Director of the United Church of Christ Commission for Racial Justice), Dr. Joseph Lowery (Southern Christian Leadership Conference), and District of Columbia Congressman, Walter Fauntroy.

At the request of Congressman Walter Fauntroy, the United States General Accounting Office (GAO) conducted a study of 8 southern states (EPA Region VIII - Alabama, Florida, Georgia, Kentucky, Mississippi, North Carolina, South Carolina, and Tennessee) to determine the correlation between the location of hazardous waste landfills and the racial and economic status of the surrounding communities. In June 1983 the report was published. The astounding results concluded that 3 out of every 4 landfills were located near predominantly minority communities.[5]

Studies on Environmental Justice

In 1987, a milestone Commission on Racial Justice (CRJ) report showed that the most significant factor in determining the siting of hazardous waste facilities nationwide was race.[6] The CRJ found that 3 out of every 5 African-Americans or Hispanics live in a community allowing unregulated toxic waste sites. By statistically correlating racial and income data from census tapes, with the location of commercial hazardous waste facilities within the same postal code, CRJ researchers concluded, "communities with greater minority percentages of the population are more likely to be the sites of commercial hazardous waste facilities." It also confirmed that "indeed, race has been a factor in the location of commercial hazardous waste facilities in the U. S." [7]

See <*http://www.readingenergy.com/publications/Justice/*> for more information. The CRJ report later became the center of examination and scrutiny because of its shocking findings. Seven years thereafter, the University of Massachusetts at Amherst published a report (1994) that found statistical flaws in the United Church of Christ CRJ report. The University of Massachusetts study determined that the CRJ's use of zip code sample sizes led to aggregation errors and conclusions that contradicted those obtained by using smaller, more accurate, census tract sample sizes. Using the more accurate sample sizes, the University researchers reported that predominantly white families tended to live closest to commercial hazardous waste facilities. It found "almost no support for the general claim of environmental inequity." [8]

The National Law Journal conducted a study of the Environmental Protection Agency (EPA) shortly after the publication of the CRJ report. The 1992 study implicated that the EPA took 20% longer to place abandoned sites in minority communities on the national priority action list and that polluters of those neighborhoods paid 54% fewer fines than polluters of white communities.[9] Since the publication of this report, the EPA created an Office of Environmental Equity in 1992, initiated a number of environmental justice projects, and took the lead in coordinating an interagency research workshop (co-sponsored by the National Institute for Environmental Health Sciences and the Agency for Toxic Substances and Disease Registry) on environmental threats in low-income and people of color communities.

Subsequent studies heightened the concern for environmental inequities and discriminatory industrial siting practices. The 1992 University of Michigan *Symposium on Race, Poverty, and the Environment* summarized that minorities bear a disproportionate burden of pollution in America. The Michigan Symposium concluded that (1) land values are cheaper in poor neighborhoods and thus attractive to "polluting industries" seeking to reduce costs, and (2) industries tend to target communities with little political clout where citizens are unorganized and lack time, money, and political knowledge[7, 10] See <*http://www.readingenergy.com/publications/Justice/*> for additional information.

Environmental Awareness

Louisiana's Petrochemical Corridor

Threatened communities in Southeast Louisiana's petrochemical corridor (the 85-mile stretch along the Mississippi River from Baton Rouge to New Orleans) typify the need for environmental awareness. Many environmentalists have dubbed the corridor "Cancer Alley." Health concerns raised by residents and grassroots activists who live in small towns along the Mississippi River have not been adequately addressed by local, state, and federal agencies. Public Health concerns finally reached the attention of the U.S. Civil Rights Commission. In its September 1993 report, the Louisiana Advisory Committee to the U.S. Commission on Civil Rights reinforce what many people already knew: African American communities along the Mississippi River chemical corridor bear a disproportionate health and environmental burden from industrial pollution created in this area.

U. S. / Mexico Border

In the late 1960s Mexico began its *maquiladora* (a "*maquiladora*" is a factory that manufactures products for other businesses) program along the United States/ Mexico border to attract foreign businesses and manufacturers. While business has boomed, the *maquiladoras* are burdening the environment with industrial toxins and pollutants. Encouraged by the North American Free Trade Agreement (NAFTA) and the devaluation of the Mexican peso in the middle to late 1990s, even more industries are rushing to take advantage of the cheap labor and lax enforcement of environmental regulations along the Mexican border. Numerous hazardous waste sites dot the border area. Heavy metals, acids, solvents, and other industrial toxins pour out of company pipes and airstacks into the surrounding communities. In addition, Mexican residents do not have access to information on *maquiladora* operations that would allow them to initiate emergency planning and protective measures.[11] See *<http://www.environmental-health.org/border.html>* for more information.

Yucca Mountain, Nevada[12]   *<http://www.umich.edu/~snre492/cases.html>*

The U. S. government has set aside Yucca Mountain, part of the Western Shoshone Nation, as a final repository for high level nuclear waste from the U.S. nuclear industry. At present, the Department of Energy (DOE) is conducting a scientific investigation of the site that will cost $63 billion and will allow for the repository to be opened by the year 2010. Although they are still investigating the area, the DOE is no longer looking for a site elsewhere. The Western Shoshone Indian tribe, who occupy nearby land (and are native to the land), are extremely concerned about observed health and environmental effects on its members. However, the federal government currently has not initiated or implemented any official health studies, remedies to the environmental pollution, programs for early detection of environmental disease, or disease surveillance programs.

Nearby communities are also being exposed to radiation from the Nevada Test Site (NTS), located on traditional Shoshone land. The U.S. and Britain have tested nuclear weapons for many years at the NTS. The Western Shoshone National Council has recognized the destruction that results from these experiments. Since 1951, approximately 1350 square miles of their 43,000 square mile territory have been destroyed by hundreds of craters and tunnels, leaving unsupervised nuclear waste sites.

There have been environmental monitoring reports issued throughout the years concerning the status of NTS, dating all the way from the 1950s to 1991. These reports prove the presence of substantial low-level radioactive releases of iodine, strontium, cesium, plutonium, and noble gases in outlying areas, with higher concentrations found in reservation communities in close proximity to NTS. Residents have reported unusual animal deaths, human hair loss, as well as an increase in cancer and birth defects.

Bhopal, India

On Monday, December 3, 1984, a toxic cloud of methyl isocyanate (MIC) gas enveloped the hundreds of shanties and huts surrounding the Union Carbide India Limited (UCIL) pesticide plant in Bhopal, India. The MIC cloud caused individuals to cough, choke, and have stinging eyes. By the time the gas cleared at dawn, many were dead or injured. Four months after the tragedy, the Indian government reported to its Parliament that 1430 people had died. In 1991, the official Indian government panel charged with tabulating deaths and injuries updated the count to more than 3800 dead and approximately 11,000 with disabilities.[13]

In the aftermath of the 3800 deaths and 400,000 injuries/illnesses caused by the leak, the Indian government further victimized the people of Bhopal. India settled out of court with Union Carbide for $470 million. That amount, distributed over nearly 550,000 claimants in the suit, did not cover the majority of medical expenses or their suffering. Upon settling, Union Carbide's stock rose $2 a share.[14]

<http://www.umich.edu/~snre492/lopatin.html>

Developing countries are often very vulnerable to exploitation by multi-national corporations. They are eager to support industrialization, but lack the infrastructure to manage it properly. Without suitable *laws* and *regulations*, developing nations are ill-prepared for such endeavors. In their efforts to attract business, these nations often (either intentionally or not) overlook the health and safety violations of the corporations doing business in their borders. Drawn by low-cost labor, new markets, and lower operation costs, corporations have little incentive to address environmental and human risks once they are entrenched.

Environmental Justice and CERCLA

Environmental justice is *equal protection under all environmental statutes and regulations*, regardless of race, ethnicity, or socioeconomic status. It includes equal participation in the decision-making process, and equal access to relief from existing environmental burdens. Executive Order 12898 was issued on February 11, 1994 to require all Federal agencies to make environmental justice part of their mission. In addition to the Executive Order, President Clinton also wrote a companion *Memorandum to the Heads of Federal Agencies* to note the role of existing federal laws in achieving environmental justice in federal programs and activities. It emphasized that "existing environmental and civil rights statutes provide many opportunities to address environmental hazards in minority communities and low-income communities."[15,16]

In spite of this, poorer neighborhoods have historically been the prime targets for many detrimental by-products of an industrial society. These communities are generally seen as paths of least resistance, and thus are more likely the targets for polluting facilities or industries. Research has shown that lead poisoning is also correlated with income. Among African-American families earning less than $6000, 68 percent of the children are lead-poisoned, compared with the 36 percent of white children of similar family income. In families with income exceeding $15,000, more than 38 percent of African-American children suffer from lead poisoning compared with 12 percent of white children. Thus, even when income is held constant, middle-class African-American children are three times more likely to be lead-poisoned than their middle-class white counterparts.[17,18]

In the area of pesticides, racial minorities, especially Latinos, are more likely to be employed as migrant farmworkers and are at increased risk of exposure to dangerous pesticides. Farm work not done by farm families is done primarily by ethnic minorities. Eighty to ninety percent of the approximately two million hired farmworkers are Latino, followed by African-Americans, Caribbeans, Puerto Ricans, Filipinos, Vietnamese, Koreans, and Jamaicans.[18,19,20] It is estimated that as many as 313,000 farm workers experience pesticide-related illnesses each year. [18,19] Not surprisingly, Hispanic women generally show higher levels of pesticides in their milk than white women.

There are nearly four million public and subsidized housing units in the U. S. that are frequently treated with toxic pesticides to exterminate insects and other pests. These units house millions of people, including a disproportionately higher number of people of color, and those who are most vulnerable to pesticide exposure are young children and elderly.[18]

The National Environmental Policy Act (NEPA) is specifically referenced in Executive Order 12898. NEPA has integrated its CERCLA response processes to include the analysis of cumulative, off-site, etiologic and socioeconomic impacts, to the extent practicable. NEPA also commits the Department of Energy to take steps to ensure opportunities for early public involvement in the CERCLA process and claims that it will make CERCLA documents available to the public as early as possible.

Each of the substantive areas identified in Executive Order 12898 emerge as potential concerns at various points in the CERCLA response processes. There are 3 major areas in the processes and they include: *(1) the distribution of impacts and benefits, (2) data collection and analysis* (including patterns of fish and wildlife consumption and multiple and cumulative effects), and *(3) public participation and access to information.*

1) Distribution of Impacts and Benefits

Among the earliest recognized and most familiar environmental justice concerns is the imposition of *disproportionate adverse impacts* on low-income and minority communities. These impacts arise from activities, such as siting new facilities or starting new projects, *including cleanups*, which have potentially adverse environmental consequences. Thus, questions emerge about whether sites in low-income and minority areas receive fair treatment in prioritization and cleanup standards.

2) Data Collection and Analysis

The conditions that give rise to environmental justice concerns are specific to individual communities and their histories. There are some exposure pathways that are more prevalent in low income communities, such as a relatively substantial reliance on fishing or hunting for food, which may lead to much higher exposures to bioaccumulating toxins than in the general population. These exposures may occur over time and have a cumulative impact. There are some low-income communities that are more likely to expose workers to unsafe working conditions, and may increase the chances for exposures from multiple sources. Additionally, the potential for overlooked hazardous waste sites is a concern.

3) Public Participation and Access to Information

CERCLA response processes require public participation at several levels. For people in low-income and minority communities, several issues may inhibit effective participation. Some reasons why public participation may not be possible or effective include *language barriers, cultural differences, long-term burdens* (such as requiring multiple response actions within low-income communities), and *the difficulty in establishing a trusting relationship with the community.* However, the need for developing a resolution to these problems will have long-term benefits for both CERCLA and the community.[16]

Removal (or corrective) actions are rapid "response cleanup" activities undertaken to *eliminate, minimize,* or *mitigate* the threat of a hazardous substance release. These removal actions are part of 40 CFR 300.415, and include:

- notification or discovery;
- site evaluation;
- action memorandum;
- response action;
- site closeout; and
- post-removal site control.

Environmental justice considerations are secondary to significant adverse health effect situations. A streamlined risk evaluation process is recommended in these instances to help determine what type of removal action will be necessary, whether available technologies can be successful in interrupting the exposure pathway, and the best way to reduce the risks associated with the removal action itself. Strategies to address community concerns during site assessment have been used by EPA at several Superfund sites. They include:

- establishing a satellite office or information repository near the site;
- distributing of fact sheets;
- conducting door-to-door surveys and results presentations; and
- holding open houses.

In the case of time-critical removal actions, staff developing the action memorandum should consider differential patterns of consumption that may lead to exposure, such as the presence of other sources of exposure to contaminants, or community history that indicates the likelihood of cumulative exposures in determining the proposed response.

Environmental Policy and Protection Rollbacks

In the mid-1990s, the advent of rising voter apathy, divisional politics, and a conservative Congress resulted in an effort to change American environmental policies. Under the guise of "regulatory reform," attempts were made to rollback environmental protections developed in the last twenty years. However, throughout 1995 and 1996, community leaders, national advocacy groups, and environmentalists worked to prevent conservative legislators from gutting existing environmental laws. Thus, there was the defeat of the so-called "regulatory reform package." [4] Additonal information is available at <*www.tsulaw.edu/environ/envhist.htm*>.

Slowly, the environmental justice message has begun to reach the U.S. Congress. Several bills have been introduced which address some aspects of environmental justice. However, state action is also needed. Arkansas and Louisiana were the first two states to pass environmental justice laws. Virginia passed a legislative resolution on environmental justice. Several other states, including California, Georgia, New York, North Carolina, and South Carolina have pending legislation to address environmental injustice.[21]

Thinking Globally

To reduce environmental degradation and pollution, nations must become unified in the search for answers. It is clear that there is widespread destruction from pollutants, irrespective of boundaries. A country may practice sound environmental protection while a neighboring one pollutes freely. Countries are finding themselves the victims of environmental destruction even though the causes of that destruction may have originated in another part of the world. Acid rain, global warming, depletion of the ozone layer, and nuclear spills and accidents are examples of governments lacking environmental responsibility.

Economic development is linked to environmental pollution even though international stock markets fail to include the true cost of pollution in its pricing of products and services; it fails to place a value on the destruction of plant and animal species, air, and water. To date, most industrialized nations have had an incentive to pollute because they did not incur the cost of producing goods and services in a nonpolluting manner. The world will pay for the true cost of the production of goods.[22]

Executive Order 12898 is a step forward, although it is only the beginning when it comes to solving environmental justice throughout the world, let alone the U. S. Implementation of the Order is unclear. Federal and state legislation is needed to strengthen the Order to significantly reduce environmental destruction. In addition, a commitment from the industrial sector, to further reduce pollution, is essential to solving environmental justice issues. Hopefully, in the next several years, global support will evolve to include environmental justice and the prevention of the destruction of the environment.

REFERENCES

1. **"Executive Summary,"** *Toward Environmental Justice*, National Academy Press, Washington, D.C., 1994.

2. **NIEHS,** National Institute of Environmental Health Sciences research programs, Department of Health, Education, and Welfare, Public Health Service, National Institutes of Health, National Institute of Environmental Health Sciences, 1994.

3. **Bunyan, B.**, Environmental Justice – Issues, Policies, and Solutions, Island Press, Washington, D.C.,1995.

4. **"Environmental Justice History,"** Thurgood Marshall School of Law - Environmental Justice Clinic, Houston, TX, 1996, http://www.tsulaw.edu.

5. **U.S. General Accounting Office**, *Siting of Hazardous Waste Landfills and their Correlation with Racial and Economic Status of Surrounding Communities,* GAO/RCED-83-168, B-211461, June 1, 1983.

6. **Commission for Racial Justice**, *Toxic Wastes and Race in the U.S.*, United Church of Christ, Commission for Racial Justice, "Toxic Wastes and Race in the U.S.," United Church of Christ, 1987.

7. **Journal of Environmental Law and Practice**, *Environmental Justice in Siting a Waste to Energy Facility: An Empowered Community of Color Sites a Waste-to-Energy Facility*, May/June, 48-53, 1995.

8. **Anderson, D. et al.**, *Hazardous Waste Facilities: Environmental Equity Issues in Metropolitan Area*s, 18 Evaluation Review 123, April 1994.

9. **National Law Journal,** *Special Issue - Unequal Protection: The Racial Divide in Environmental Law,* September 21, 1992.

10. **Mohai, P. and Bryant, B.**, *Environmental Racism: Reviewing the Evidence*, University of Michigan Law School Symposium on Race, Poverty, and the Environment, Ann Arbor, MI, January 23, 1992.

11. **Environmental Health Coalition**, *Border Environmental Justice Campaign,* San Diego, CA, 1998.

12. **Kendziuk, J.**, *Environmental Justice Case Study: The Yucca Mountain*

High-Level Nuclear Waste Repository and the Western Shoshone, University of Michigan School of Natural Resources and Environment, 1999, http://www.umich.edu/~snre492/cases.html.

13. **Browning, J.** *Crisis Response: Inside Stories on Managing Under Siege*, Gottschalk, J., Ed., Visible Ink Press, Detroit, MI, 1993.

14. **Lopatin , J.**, *Environmental Justice Case Study: Union Carbide Gas Release in Bhopal, India*, University of Michigan School of Natural Resources and Environment, 1999, http://www.umich.edu/~snre492/lopatin.html.

15. **Clinton, W.,** *Memorandum to the Heads of Federal Agencies,* President William J. Clinton, February 11, 1994.

16. **DOE, Department of Energy,** *Incorporating Environmental Justice Principles into the CERCLA Process*, U.S. Department of Energy, Office of Minority Economic Impact, DOE/EH-413 9812, May 1998.

17. **Anonymous,** See the Agency for Toxic Substances and Disease Registry, 1988.

18. **Wright, B.,** *Environmental Equity Justice Centers: A Response to Inequity*, in Environmental Justice – Issues, Policies, and Solutions, Bryant, B. Ed., Island Press, Washington, D.C., 1995.

19. **Wasserstrom, R. and Wiles, R.**, *Field Duty, U.S. Farm Workers and Pesticide Safety,* Study 3, World Resources Institute, Center for Policy Research, Washington, D.C., 1985.

20. **Moses, M.,** *Pesticide Related Health Problems and Farmworkers*, *Am. Assoc. of Occup. Health Nurses J.*, 37, 115-130, 1989.

21. **Anonymous,** T*he Environmental Justice Movement: Continuing the Struggle for Civil Rights*, (sub)TEX, Vol. 1, 5, University of Texas at Austin, TX, February 1995.

22. **Bunyan, B.,** *Environmental Justice – Issues, Policies, and Solutions,* Island Press, Washington, D.C., 1995.

WEB LINKS FOR THIS CHAPTER

Beginnings of the Environmental Justice Movement
http://www.tsulaw.edu/environ/envhist.htm

Environmental Policy and Protection Rollbacks
<www.tsulaw.edu/environ/envhist.htm>

Manufacturing Along the Mexican Border
*http://www.environmentalhealth.org/border.htm*l

Summary of Executive Order 12898
http://www.nap.edu

Union Carbide and The Bhopal, India Disaster
http://www.umich.edu/~snre492/lopatin.html

Yucca Mountain and Nuclear Waste
http://www.umich.edu/~snre492/cases.html

ISO
14000

14

David Gillum

INTERNATIONAL ENVIRONMENTAL COMPLIANCE

With an international marketplace established through new and existing trade agreements between nations, a provocative environmental strategy has been developed to control and prevent ecological abuse of global resources. Recent legislation has resulted in increased liability for pollution, waste disposal. and toxic releases by any company. In addition, the prices for natural resources and energy have risen, while more government imposed restrictions continue to be written. It is for these reasons that more industries have an incentive to pay closer attention to health, safety, and environmental issues.[1] *<http://www.latinsynergy.org/iso.htm>*

In June 1992, the United Nations Conference on the Environment and Development (UNCED) was held in Rio de Janeiro, Brazil. The conference was attended by over one hundred countries, each agreeing on the need for further development of international environmental management programs. The conference placed a worldwide focus on corporate environmental management.[2] Equally important was identifying the number and extent of regional, national, and state environmental management schemes that included labeling and auditing. The acceptance of ISO 9000 throughout the world formed a natural paradigm from which to develop international environmental management standards. The UNCED petitioned the International Organization for Standardization to consider developing environmental management standards.[2]

WHAT IS ISO?

The International Organization for Standardization (ISO) is a private sector, international standards body based in Geneva, Switzerland. The short form "ISO" is not an acronym, but instead is derived from the Greek word "*iso*," meaning "equal" (implying "standard").[3] Founded in 1947, ISO promotes the international harmonization and development of manufacturing, product and communications standards. ISO has promulgated more than 8000 internationally accepted standards ranging from paper sizes to film speeds. More than 120 countries belong to ISO as full voting members, while several other countries serve as observer members. The U. S. is a full voting member and is officially represented by the American National Standards Institute (ANSI). ANSI is a non-governmental, non-profit organization that promulgates standards.

All standards developed by the International Organization for Standardization are voluntary. There is not a legal requirement forcing a country to adopt them. However, some countries and industries often adopt ISO standards as national standards. The government of a particular country may adopt these standards for mandatory environmental compliance.

ISO created a Strategic Advisory Group on the Environment (**SAGE**) in August 1991 to assess the need for international environmental management standards and to recommend an overall strategic plan for such standards. SAGE also recognized the potential for diverse regional, national, and state environmental standards that may result in unintended technical barriers to international trade and commerce. Thus, the need for *voluntary international standards* for environmental management systems was immediately acknowledged. SAGE was asked to consider whether environmental management standards could serve the following:

1. Promote a common approach to environmental management similar to ISO 9000 Quality Management Standards;
2. Enhance an organization's ability to attain and measure environmental performance; and
3. Facilitate trade and remove trade barriers.

Upon conclusion of the study, SAGE recommended that an ISO technical committee formally consider and produce final "consensus" standards. Thus, in January 1993, the Technical Committee 207 (TC-207) was created. TC-207 is now a worldwide delegation consisting of standard writing committees. TC-207 was divided into subcommittees, which were further subdivided into Working Groups where the standards are actually written (Figure 15.1),[4] <*http://www.epa.gov/epaoswer/non-hw/industd/guide/app8-9.pdf*>.

Canada was awarded the secretariat for TC-207 and the Inaugural Plenary Session was held in Toronto, Canada, on June 2 and 3, 1993. Over 200 delegates from over 30 countries and organizations were in attendance. The TC-207 framework also includes coordination and cooperation with regional and international governmental bodies including the European Union (EU), the General Agreement on Tariffs and Trade (GATT), and the Organization for Economic Cooperation and Development (OECD). TC-207 meets annually to review the progress of its subcommittees. [2]

TC-207 consists of 6 subcommittees, one for each focus area, and a working group:

> **SC1:** *Environmental Management Systems* (Secretariat – United Kingdom; administered by the British Standards Organization)
> **SC2:** *Environmental Auditing and Related Environmental Investigations* (Secretariat – Netherlands; administered by the Nederlands Normalisatie Institüt)
> **SC3:** *Environmental Labeling* (Secretariat – Australia; administered by Standards Australia)

SC4: *Environmental Performance Evaluation* (Secretariat – U.S. administered by the American National Standards Institute)

SC5: *Life-Cycles Assessment* (Secretariat – France; administered by Association Francaise de Normalisation)

SC6: *Terms and Definitions* (Secretariat – Norway; administered by Norges Standardiseringsforbund)

WG1: *Environmental Aspects in Product Standards* (Secretariat – Germany; administered by the Deutsches Institüt fur Normünge. V.)

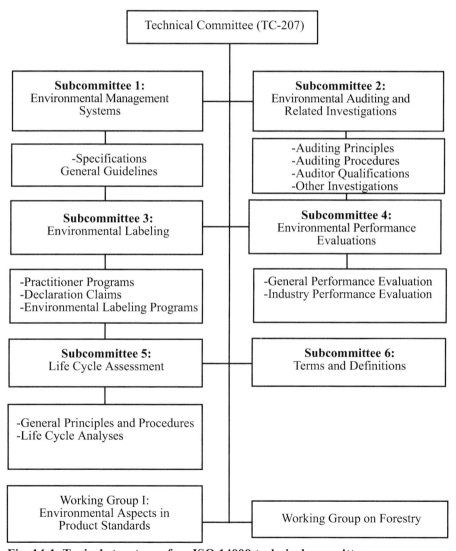

Fig. 14-1. Typical structure of an ISO 14000 technical committee.

Voluntary Management

An Environmental Management System (EMS) is a voluntary program established by businesses and organizations that results in the integrated management of environmental practices and prevention of noncompliance with environmental regulations. These programs consist of a company's overall environmental policy, the safeguards developed and implemented to prevent non-compliance, and the regular procedures (including internal or external compliance and management audits) to evaluate, detect, prevent, and remedy any environmental problems associated with the institution's activities,[5] *<http://web.ansi.org/public/iso14000/subject/manage.html>*.

ISO 14000 Environmental Management Series

The ISO 14000 series (ISO 14001 – ISO 14050) is a family of environmental management standards designed to provide an internationally recognized framework for *environmental management, measurement, evaluation,* and *auditing.* It is the first set of standards ever established in consultation with the global manufacturing community. The ISO 14000 series provides organizations with the tools to assess and control the environmental impact of their activities, products, or services. The ISO 14000 series is designed to be flexible enough to be used by any organization of any size, in any field. They address the following subjects:,[6] *<http://www.iso.ch/9000e/iso14000.pdf>*.

1. Environmental management systems (ISO 14001) - the overall environmental policy to assure compliance, including all procedures involved
2. Environmental auditing and guidelines (ISO 14010–14012) - assessments, surveys, reviews, or appraisals of environmental management systems
3. Environmental performance evaluation (ISO 14031) - an internal evaluation system providing consistent information to support management decisions regarding the organization's relationship with the environment
4. Environmental labeling (ISO 14020–14025)- provides information about a product or service in terms of its overall environmental character, a specific environmental attribute, or any number or attributes
5. Life-cycle assessment (ISO 14040–14043) - a management tool encompassing the assessment of environmental impacts of a product in its entire life cycle;
6. Environmental aspects in product standards (ISO Guide 64) - incorporating environmental aspects or considerations into the development of standards,
7. Terms and definitions (ISO 14050) - a comprehensive list of terms and definitions for environmental management.

The standard for environmental management systems, ISO 14001, is the only one to which an organization may be **certified**. The Guidance document *(ISO 14004)*, the Environmental Management Systems standard *(ISO 14001)*, and the auditing standards *(ISO 14010, 14011, and 14012)* have been operational since 1996.

The ISO 14000 series continually emphasizes the need to recognize the importance of environmental protection and a commitment to such protection. The series' key principles require recognition that:

1. Environmental management is among the highest priorities;
2. Dialogue with internal and external interested parties be established;
3. Management and employee commitment to the protection of the environment be developed with clear assignment of accountability and responsibility;
4. Environmental strategic planning throughout the product or process life-cycle be encouraged;
5. Appropriate and sufficient resources, including employee training, be provided to achieve targeted performance levels on an on-going basis; and
6. Environmental performance be assessed against appropriate policies, objectives, and targets.

More importantly, the ISO 14000 series is a way of protecting the environment through increased pressures on organizations to perform in an environmentally sound manner. [2]

The emphasis behind an EMS is one of prevention rather than corrective action; a proactive stance rather than "fire-fighting." In addition, the ISO 14000 series provides a basic structure for industrial firms to improve their environmental performance through the establishment of environmental goals, implementation of a plan for achieving those goals, monitoring progress, and corrective action. The ISO 14000 series is a tool enabling organizations to incorporate environmental elements into their management system.

How does an environmental management system (EMS) work? ISO 14001 is an EMS that is structured to be applicable to virtually any industry. An EMS is a management structure that allows an organization to assess and control the environmental impact of its activities, products, or services.

The ISO 14000 series was designed to accommodate a wide range of business sizes and types, and as such, varying uses of the ISO 14000 series are possible. ISO 14001 "registration" is the formal recognition of an organization's EMS. Some organizations, mainly those outside North America, refer to this as "certification." Organizations may simply declare that their EMS meets the requirements of ISO 14001 ("self-declaration"). However, many organizations choose to have their EMS registered, usually to provide greater assurance to clients or the public, or because regulators or clients require it. An independent third party, known as a "registrar," assesses and audits the organization's EMS to ensure that it complies with the requirements of the standard. The primary benefit of certification is validation by an outside agency that the implemented EMS meets the requirements of the ISO 14000 standard.

A further level of confidence is provided by the accreditation of registrars. Accreditation is the evaluation of an organization's competence to carry out certain functions. It provides the basis for national and international acceptance of a registrar's certificate of registration.

To become ISO 14000 certified, an organization must:

1. Develop an environmental policy;

2. Establish procedures to identify environmental aspects of all activities in order to determine which have significant impact;
3. Document environmental objectives and targets;
4. Include a commitment to the prevention of pollution;
5. Train all personnel whose work may significantly affect the environment;
6. Create an auditing system to ensure the program is properly maintained; and
7. Review the audit results to ensure continual improvement.

The standard does not require organizations to meet top-level performance standards but instead standardizes the field of environmental management tools and systems. The following are its core principles:[2]

1. Commitment to environmental management (one of the highest corporate priorities) should permeate the organization through all levels;
2. Identification of legislative and regulatory requirements and environmental aspects associated with the organization's activities, products, and services;
3. Development of a management process for achieving objectives and targets;
4. Providing appropriate financial and human resources to achieve objectives and targets;
5. Assignment of clear procedures for accountability and responsibility;
6. Establishment of a management review and audit process to identify areas for improvement; and
7. Development and maintenance of communication with internal and external interested parties.

An organization's ability to conform to the ISO 14000 standards is assessed through the EMS, which consists of 5 main paragraphs, each with subsections. The 5 main sections are *environmental policy, planning, implementation and operation, checking and corrective actions, and management review.* A brief summary for each is listed: [2]

1.0 Environmental Policy

Environmental policy is defined as the "Statement by an organization indicating its intentions and principles, in relation to its overall environmental performance, which provides a framework for action and for the setting of its environmental objectives and targets."[7] *<http://www.iso.ch/markete/more14k.htm>* Details of the environmental policy shall be included.

2.0 Planning

This section provides a process by which organizations can identify significant *environmental aspects* that should be addressed as priorities. Environmental aspects are the elements of an organization's activities, products, and services, which can interact with the environment.

3.0 Implementation and Operation

This section includes the identification of the responsible party for the EMS requirements as well as communication and documentation of these roles and responsibilities at each level and/or function. Resources must be available to support an effective EMS, including human as well as financial and technical resources.

4.0 Checking and Corrective Action

A procedure must be maintained in which all key elements of operations and activities that can have a significant impact on the environment are monitored and measured. This includes recording information to track performance, relevant operational controls, and conformance with objectives and targets.

5.0 Management Review

Senior managers must review the EMS for suitability, adequacy, and effectiveness at appropriate intervals to ensure continual improvement. An organization is a dynamic entity, thus changes in the EMS policy objectives or other elements may be necessary.

What are the benefits of ISO 14001?

The ISO 14000 series has been labeled a "green passport" leading to a fast "trust track" between an organization and its regulatory body. An EMS effectively incorporated demonstrates to an interested party that environmental policies, objectives, and expectations have been met; that emphasis is placed on prevention rather than corrective action; that a company can provide further evidence of reasonable care and regulatory compliance; and that the organization supports continual improvement. An effective EMS provides an organization with the ability to balance and integrate economic and environmental interests. It also describes a number of benefits that may translate into a financial advantage.

An ISO 14001 EMS gives a company the tools to monitor and improve the impact on the environment. As a result, having one may help in the following ways: [2,8]

1. obtain insurance at reasonable cost
2. enhance image and market share
3. meet vendor certification criteria
4. ability to dispose of waste
5. liability limitation and risk reduction
6. demonstration of due diligence
7. save in consumption of energy and materials
8. easier site selection and permitting
9. technology development and transfer
10. improve industry-government relations
11. meet customer's environmental expectations
12. maintain good public/community relations
13. satisfy investor criteria

14. reduce cost of waste management
15. lower distribution costs
16. provide a framework for continuous improvement of environmental performance

In addition, proponents of using the ISO 14000 series as a basis for demonstrating good-faith compliance actions foresee such possibilities as an ISO 14001 certified organization having fewer inspections, fast-track permitting, and mitigation and/or reduction of the severity of penalties for violations.

Examples of Compliance

Companies spend an estimated $500 billion globally on environmental protection. In the U.S. market alone, annual environment-related spending is about $250 billion. Experts predicted at the current rate of increase that environmental spending will have exceeded 3% of the U.S. Gross National Product after the year 2000. The Chemical Manufacturers' Association (CMA) estimates that environmental compliance-related requirements, in addition to remediation spending, now average 15% of capital and 5-10% of operating expenses. [9]

Ensuring safety and environmental protection through an effective management system generally makes good business sense. The consequences of an oil spill,[2] for example, can be significant: Crude oil may be purchased for approximately 40 cents per gallon; if it were spilled, the "total spill unit value" defined as the total value for cleanup, third-party damages and natural resource damages, averages $245 per gallon. An effective EMS may result in reduction of a company's risk of fines, and/or litigation costs by timely correcting of non-compliance. It may also change pollution control into an everyday *activity* rather than a *reaction* during a crisis.[2]

Companies and industries are discovering that environmentally friendly products, as well as placing an effort in reducing waste, will increase marketplace appeal and allow more customers to achieve their own environmental objectives. McDonald's restaurant corporation discovered this lesson during the 1980s after it was criticized for the large quantities of packaging used in its restaurants. McDonald's agreed to implement aggressive efforts to redesign its packaging as part of a legal settlement with the Environmental Defense Fund. The fast food company eliminated more than 10,000,000 pounds per year of solid waste generated. The company then identified another 35 to 50 million pounds of additional waste that could be reduced. Not only did McDonalds reduce their waste production, they also appeared proactive in supporting the environment.[9]

THINKING GLOBALLY

Since adoption of the standard in 1996, over 5000 organizations worldwide have received ISO 14001 certification, with the number increasing daily. Currently only about 3% of the certified organizations are U.S. companies, but interest is expected to grow due to pressure from international trading partners, regulatory agencies, and the public. The U.S. EPA and a number of states formed the Multi-State Working Group (MSWG) on Environmental Management Systems to determine whether the use of an ISO 14001 EMS has a positive effect on an organization's environmental performance, including regulatory compliance,[10] *<http://www.naep.org/NAEP/Conference/1999/Abstracts/tmp/old/author/Edwards.html>*.

Seven major trade associations and industries (including the General Electric Company; American Automobile Manufacturers Association; American Forest and Paper Association; American Iron and Steel Institute; American Petroleum Institute; Chemical Manufacturers Association; Edison Electric Institute; and Electronic Industries Alliance) have formed a broad-based industry coalition to address public policy issues associated with the implementation of ISO 14000. The coalition insists that ISO 14000 should remain voluntary and resist any governmental efforts to make it part of the regulatory process. The group claims that the "ISO 14000 standards could have an important impact (positive or negative) on the effective implementation of the standards." In contrast, on March 12, 1998 the U.S. EPA issued a document that states that "Implementation of an EMS has the potential to improve an organization's environmental performance and compliance with regulatory requirements,"[5] *<http://web.ansi.org/public/iso14000/news/09-97emr_17.html>*.

The ISO 14000 standards offer a new approach to protecting the environment. They are aimed at improving industrial environmental performance and easing the burden of environmental regulation. The systematic approach to ISO 14000 may prove effective in encouraging organizations to take a proactive, preventative, and holistic approach to managing their environmental impacts. The daunting task of achieving *full-compliance*, though vigorously pursued, remains elusive. Thus, some believe it is important that environmental policy and environmental laws encourage voluntary compliance, self-policing, and audits in which diligent correction of non-compliance is not penalized.[2]

There is fear that compliance with ISO 14000 will not guarantee environmental improvements.[11] *<http://www.p2.org/iso.html>* In addition, ISO 14000 standards are not designed to aid enforcement of environmental laws or inform the public of problems. Thus, the determination as to whether or not ISO 14000 is effective will probably not be decided until it is widely accepted and implemented on a global level.

REFERENCES

1. **Coates, R.,** "ISO 14000 - The World Adopts Environmental Standards," The Latin American Alliance, 1997. http://www.latinsynergy.org/iso.htm

2. **von Zharen, W. M.**, ISO 14000 - Understanding the Environmental Standards, "Development of International Standards," Government Institutes, Inc., Washington, D. C., 1996.

3. **Webster's Encyclopedic Unabridged Dictionary of the English Language**, Gramercy Books, 1989.

4.   **USEPA,** "ISO 14000 Resource Directory," EPA/625/R-97/003, Contract #68-3-0315, United States Environmental Protection Agency, Office of Research and Development, Office of Pollution Prevention and Toxics, October 1997. http://www.epa.gov/epaoswer/non-hw/industd/guide/app8-9.pdf

5. **ANSI Online**, "ISO 14000 - Environmental Management Systems," November 17, 1997. http://web.ansi.org/public/iso14000/subject/manage.html

6. **International Organization of Standardization**, "ISO 14000 - Meet the Whole Family," 1998. http://www.iso.ch/9000e/iso14000.pdf

7. **USEPA,** "ISO 14001," Section 3.10, ISO 14000: International Environmental Management Standards, Office of Research and Development, Office of Pollution Prevention and Toxics, Washington, D.C., 1998.

8. **International Organization of Standardization**, "ISO Standards Compendium: ISO 14000 - Environmental Management," 1998. http://www.iso.ch/markete/more14k.htm

9. **Nestel, G.**, "The International Environmental Management Standard," in ISO 9000: A Comprehensive Guide, 478, 1997.

10. **Edwards, B., Gravender, J., Killmer, A., Schenke, G., Tsukiji, H., and Willis, M.,** "Assessing the Effectiveness of ISO 14000," Donald Bren School of Environmental Science & Management, University of California, Santa Barbara, CA, 1999.

http://www.naep.org/NAEP/Conference/1999/Abstracts/tmp/old/author/
Edwards.html

11. **Pringle, J., Leuteritz, K., and Fitzgerald, M.**, "ISO 14001: A Discussion of
    Implications for Pollution Prevention," National Pollution Prevention
    Roundtable - ISO 14000 Workgroup, White Paper, January 28, 1998.
    http://www.p2.org/iso.html

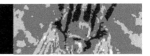

WEB LINKS FOR THIS CHAPTER

Environmental Management System (EMS)
http://web.ansi.org/public/iso14000/subject/manage.html>

Environmental policy, objectives, and targets
http://www.iso.ch/markete/more14k.htm

Implementation of an EMS has the potential to improve an organization's environmental performance and compliance with regulatory requirements
http://web.ansi.org/public/iso14000/news/09-97emr_17.html

Industries dealing in health, safety, and environmental issues
http://www.latinsynergy.org/iso.htm

Multi-State Working Group (MSWG) on Environmental Management Systems
*http://www.naep.org/NAEP/Conference/1999/Abstracts/tmp/old/author/Edwards.ht
ml*

Subjects addressed by ISO 14000
http://www.iso.ch/9000e/iso14000.pdf

Compliance with ISO 14000
http://www.p2.org/iso.html

Working Groups of ISO
http://www.epa.gov/epaoswer/non-hw/industd/guide/app8-9.pdf

LIABILITY AND ENFORCEMENT 15

Andrea Depatie and Kevin Doherty

Environmental Enforcement and Liability are important issues in preserving our environment. Environmental laws and statutes are written and enacted with the intention of bettering the atmosphere and human health. However, simply because a law is passed doesn't mean facilities and polluters will automatically change their procedures. The theory of compliance offers a method that provides a system to assist laws in producing positive results.[1] The theory of compliance consists of four parts: implementation, compliance, enforcement, and effectiveness. Following these steps and holding responsible parties liable is expected to motivate companies and industries to meet the requirements for creating a positive change in the condition of the environment.

Implementation refers to the legislative, organizational, and practical actions that are taken to make a law operative.[1] When a law becomes implemented it is punishable if it is not followed. Compliance is conforming to the law and following the written statutes and requirements. Failure to comply with the law will result in criminal sanctions. Enforcement is the method that is available to individuals (enforcement agents, corporations, and citizens) to ensure that the law is being followed properly and safely.[1] Examples of the types of enforcement available are: testing and monitoring, giving or withholding funds, civil damages, fines, denial of certification, and injunctions.[2] Lastly, effectiveness is the degree to which individuals and corporations have met a law's objectives and goals. The ultimate level of effectiveness is many times related to the amount of enforcement exerted.[1] This demonstrates the need and importance of enforcement in this process.

With each year, new methods and information are being discovered about environmental issues. The attitude of compliance is also changing. Today, many incentives are being offered to companies that follow the law. This new incentive approach differs from the traditional command and control approach. The command and control approach is dependent upon someone violating the laws followed by a punishment. The new incentive approach offers companies more creative ways of changing their methods, creating a lot less work for enforcement agents. This also frees the time of enforcement agents to detect other violations, monitor those guilty of the violations, and provide more accurate and effective information at a quicker rate.[1]

Agents have been more stringent in prosecuting violators through the Department of Justice (DOJ) because in many situations companies are now offered more flexible methods in correcting the violations. The Department of Justice is responsible for handling all of the violations detected by enforcement agents.[2] The Environmental Protection Agency (EPA) cannot directly handle or prosecute any of their findings without the legal proceedings of the Department of Justice. The Department of Justice contains the Organization of Environment and Natural Resource Division. The purpose of this division is to hold the guilty parties of environmental violations responsible for the costs of cleanup and fines instead of spending the public's tax dollars. Within the Division of Environment and Natural Resource there is nine offices, including the Environmental Crimes Section, the Environmental Defense Section, and the Environmental Enforcement Section <*ww.usdoj.gov*>. The individual descriptions of these sections can be found at <*http:// www. usdoj.gov/enrd/enrd-home.html*>.

The EPA and the DOJ fight and prosecute environmental crime. In order to be considered an environmental violation an action must break an environmental law and endanger a person and/or the environment. Offenders are usually corporations or individuals.[2] The EPA can find individuals or companies guilty of a criminal violation or a civil violation. Criminal violations tend to take a great amount of time and money to resolve. Many times they also result in a punishment less strict than originally proposed.[3] Civil violations usually result in fines or corrective action of the violation. These civil proceedings are usually easier to prove and result in less lengthy court proceedings.[3]

There are three kinds of criminal violations that are written into laws and statutes. They are defined as negligent actions, knowing violations, and knowing endangerment.[3]

Negligent action. This is an act that is not intended to inflict injury. However, negligent acts are also handled as a criminal case if it can be proven there was intent to commit an injury. Under the definition of negligence there are three main factors that include:

1. the committing or omitting of a negligent act,
2. owing another by legal duty imposed under the law, or
3. causing or resulting harm to the plaintiff or the plaintiff's property.[4]

In negligence cases a defendant will be liable if the plaintiff proves the defendant has a duty to prevent or to avoid harm to the plaintiff, that the defendant failed to meet the requirement, and that the failure was the proximate cause of the harm suffered to the defendant.[5] The proximate cause of injury is known to be the natural and continuous sequence of actions, unbroken by intervention, which causes injury. Without those actions the ending result of injury would not have occurred.

If an individual is harmed by a negligent action, such as the improper disposal of a hazardous waste, they can file a suit in order to get reimbursed for their losses. In such a case, a defendant cannot defend him or herself by stating that they had full compliance with all government regulations and that they had all the necessary permits. In some states an individual can be held liable, without the need for further evidence, if they did not comply with the regulations or attain the needed permits.

Knowing violations: A knowing violation is the second kind of criminal violation that is written into law. Knowing violations cover three different actions:

1. knowing of one's actions as opposed to an accidental occurrence;
2. if action is taken to avoid the truth, knowledge can be inferred; and
3. claiming ignorance of the law since knowledge can be inferred by one's position in the company.[5]

Any employee that participates directly in an illegal act is subject to criminal provisions. Also included is any employee who did not personally participate in or have knowledge of illegal activities, but who are in certain managerial positions or work in an area governed by health and safety statutes. We should in fact have such knowledge and so may be charged with the violation. The manager will be held liable if:

1. the violation occurred within the area of the manager's supervision or control,

2. the manager had the authority and power to prevent the act,

3. the manager knowingly did not correct the action. [4]

Criminal liability will be passed onto the manager who is responsible for supervising and directing employee activities that are regulated by environmental laws.

This doctrine of implying liability upon the manager or operating officer is being used more today. In the case *U. S. v. MacDonald and Watson Waste Oil* (1991) the court found the executives guilty of the violations committed stating that there was a sufficient amount of evidence to demonstrate that they had knowledge of the acts, although they were not physically involved.[2]

Knowing endangerment: The third type of criminal violation is knowing endangerment. In a knowing endangerment case, the court must try to determine whether the defendant knew that a violation placed another person in imminent danger of death or serious bodily injury. The court is required to demonstrate the defendant's knowledge in order to proceed with prosecution. A manager will be held liable of knowing endangerment if he or she ignores, refuses to act on, or attempts to avoid information of illegal activity.[2]

THE U. S. COURT SYSTEM

The U. S. Court System will ultimately decide the sentence of the accused individuals. This is an important factor in determining how society will respond to the enforcement agents and the regulations they have violated. For example, if the court system gave out verbal warnings to everyone found guilty of violating the law, there would not be a lot of motivation to correct their negative actions. However, if facility managers and individuals are fined or even imprisoned, corrective methods would be found at a quicker rate. So, it is important that the courts uphold the law in the best manner possible for the environment and for human health.

In the 1980s, Congress amended and improved environmental legislation that strengthened the criminal penalties that are available to punish violators.[3] Congress focused on the Resource Conservation and Recovery Act (RCRA), making it a felony for a person to treat, store, or dispose of hazardous waste without a permit.[3] This amendment is the first felony sanction incorporated into environmental legislation.[3]

Further revisions have been made on RCRA, making violators of this act punishable by a prison term up to 5 years.[2] Following RCRA's lead, other environmental legislation was reviewed and amended with criminal punishments written into them.

The Department of Justice (DOJ), the EPA, and the Federal Bureau of Investigation (FBI) in the 1980s began prosecuting violators and the officers of corporations because of the stringent punishments in the legislature. In 1983, the DOJ's enforcement budget was $257,000.[3] By 1993, the enforcement budget was $44.5 million.[3] This shows how important the aspect of enforcement is in preserving the law. In 1990, the Department of Justice handled 134 criminal indictments.[3]

The U. S. courts also play an important role in protecting our environment and prosecuting violators of the law. The U. S. Court System is active at the state and local level as well as the federal level. State and local courts differ from state to state in terms of their regulations and laws. State courts also differ from the federal court system. For example, Illinois imposes a fine of $500,000 per day and up to 7 years of imprisonment if found guilty of criminal disposal of hazardous waste.[3] The federal Resource Conservation and Recovery Act punishes individuals for committing the same violation of disposing of hazardous waste by a $50,000 fine and/or 5 years in prison.[3] This demonstrates the differences in the state and federal level and that a state may enforce stronger punishments than the federal government. It is common for state legislation to vary from state to state and from the federal law. Therefore, it is important to know the law of the state, in which you are working.[2] Many times facility managers and operators are held liable for regulations that are different from the ones they knew, because they are working in a different state.[2] Many of the violators will be dealt with at the local or state level. The local courts have the power to handle most of the cases they are confronted with. However, if one of the parties is dissatisfied with the outcome or an appeal is filed, the case may move into the court of appeals. There are twelve courts of appeals in the U. S.[2] If after a case is heard at the court of appeals, and one or both of the parties are still dissatisfied, a petition may be granted to send the case to the Supreme Court. However, just because a petition is filed for the U. S. Supreme Court, does not mean that it will be heard. The Supreme Court is highly selective in what cases they will hear.

The stratified organization of our court system presents a sound opportunity for a fair trial. Any errors or discrepancies made at any level of the court system may be overruled or retried.[2] This represents a system of checks and balances in an attempt to provide fairness in the court system. However, a lot of individuals: judges, lawyers, witnesses, and jurors are responsible for deciding the fate of the accused, and may times differing views and perspectives come into play.[5] The involvement of many individuals has the potential to create mistrials and hung juries.

The Supreme Court of the U. S. handles cases that are of national importance. Being the highest level of court available in the U. S., the judges that serve on the panel are appointed by the president. Supreme Court judges serve for life on the panel once they are appointed. This can create positive and negative impacts on the cases being tried, depending on the issue. For example, during the years that President Reagan served he developed a reputation for being an "anti-environment" president.[1] He appointed Supreme Court judges that reflected his same beliefs and ideas. Therefore, many times environmental legislation would not pass through the Supreme Court during his reign.

In the 1960's, the civil rights movement was alive in the U. S. People who wanted to gain their rights during this time began using the court system as a way of being heard and as a way of advancing their cause. Today, this practice still continues. In most of the environmental legislation enacted, there are clauses that allow private citizens to sue the agency in violation of the law. The private citizen has to believe that the agency is not performing in accordance to the law and many times has suffered as a direct action of the agency's negligence.

Sentencing by the Courts

Regulatory agencies have several enforcement options available to them. If a violation has occurred, regulatory agencies will prosecute the highest-ranking individual they can within the company. Many times this is the facility manager or operator.[4] In order to protect their reputation as well as their company's they should be familiarwith the laws and the consequences of being found guilty of a violation. Regulatory agencies can propose the following regulatory options on those found guilty of a violation:

1. Do nothing;
2. Widespread negative publicity against the company or violator;
3. Revocation of existing permits and refusal to renew expiring permits;
4. Administrative penalty proceedings involving smaller penalties and corrective action;
5. Civil actions for monetary penalties and injunctive relief;
6. Criminal prosecution of the violator and sometimes of responsible management personnel.

When an individual is found guilty of a violation, a sentence is proposed in accordance to the action. One of the most common criminal violations is providing false information or statements. This action is punishable by a fine, 5 years of imprisonment, or both.[4]
Recent amendments to Title 18 have allowed prosecutors to increase the

amount of fines that can impede in environmental cases.[4] Courts are given there sentencing power under their provisions 44 18 U.S.C.§571, which allow fines for the set violation or the following options:

1. Felony: $250,000/individual and $500,000/corporation
2. Misdemeanor resulting in death: $250,000/individual and $500,000/corporation
3. Misdemeanor: $5,000-$100,000/individual and $10,000-$100,000/corporation

In an effort to quickly remedy environmental violations, fines accumulate on a daily basis. Until February 1987, fines of $10,000 a day were enforced; they were then increased to $25,000 a day for certain violations.[3]

REFERENCES

1. **Vig, N.J. and Axelford, R.S.,** *The global environment: institutions, law and policy.* Congressional Quarterly Inc.,Washington, D.C., 1, 137, 1999.

2. **Situ, Y. and Emmons, D.,** *Environmental crime: the criminal justice system's role in protecting the environment.* Sage Publishers, Thousand Oaks, CA., 1, 124, 2000.

3. **Reinke, D.C. and Swartz, L.L.,** *The NEPA reference guides,* Battelle Press, Columbus, OH., 1, 206, 1999.

4. **Wing, K.R.,** *The law and the public's health.* Health Administration Press, Chicago, IL., 4, 9, 1995.

5. **Dowden, L. and McNurney, J.,** *Municipal environmental compliance manual.* Lewis Publishers, Boca Raton, FL., 1, 23, 1995.

6. **U. S. Department of Justice**
   http://www.usdoj.gov/enrd/enrd-home.html

WEB LINKS FOR THIS CHAPTER

Compliance Online
http://www.ieti.com/taylor/compliance.html

Environmental Compliance and Technology On-Line
http://www.eiccd.cc.ia.us/ecat/index.html

EPA Compliance and Enforcement
http://es.epa.gov/oeca/index.html

Environmental Protection Agency Compliance Resources
http://www.federalregister.com/hpages/envlink.html

The Environmental Compliance Assistance Center
http://www.hazmat.frcc.cccoes.edu/homepage.htm

INDEX

A